Illustrations: Author

Other Titles By Author

Fiction
1939
Citizen Robot
Amsterdam in Ultramarine
Ghosts At War

Acknowledgments

This book would not exist without the patience, encouragement, and good judgment of my wife, **Jaana Narsipur**, who listened to more opinions about refrigerators and dark patterns than any reasonable person should be asked to endure.

To **Steve Krug**, whose four words started a conversation that this book attempts to continue twenty-five years later — thank you for writing the first half of my title.

And to **Sabrina Demming** — wherever you are — I'm sorry I doubted you.

DON'T MAKE ME THINK, BUT DON'T THINK FOR ME

The Joys and Horrors of AI Design

By: Andy Grogan

Table of Contents

Introduction: The Dumbest in the Room

This is a book about technology and the people who use it. More specifically, it is a book about the gap between those two things — the growing, consequential, occasionally infuriating gap between what technology does and what the people using it **actually need**. It is about artificial intelligence, and automation, and the design decisions that determine whether those things make your life better or quietly, persistently worse.

But before we get into any of that, I should tell you who I am.

I am the Dumbest in the Room.

This is not false modesty. It is a job title, self-assigned, and — according to my wife, who has observed the phenomenon at close range — something I appear to have a natural talent for. I have practiced it in boardrooms and design studios, in enterprise software companies and scrappy nonprofits, in New York and Amsterdam and various conference rooms in between. I claim the title early, usually during introductions, before the meeting has found its rhythm and the usual scramble to establish who is the smartest person present has gotten properly underway.
The scramble, in my experience, is almost always counterproductive. The smartest person in the room is rarely the most useful one. The most *useful* person in the room is the one paying attention to the human being who isn't there.

That is what the Dumbest in the Room does. That is, at its core, what UX design does.
Let me introduce you to two people, who live on opposite sides of the seam between technology and people.

The Person Who Isn't in the Room

She is on the sofa. Let's call her Sofia. It is nine o'clock on a weeknight. The kids are finally in bed — this took longer than expected, as it always does. The dog is campaigning for attention with the focused determination of an experienced lobbyist. The television is on, something she has been meaning to watch for three weeks and is now half-watching while her laptop is open.

She needs to do something. Book a flight, maybe, or check her insurance, or order something that should have been ordered last week. The task is not the point of her evening. It is an interruption — a necessary one, but an interruption nonetheless. She would like to complete it with the minimum possible effort and get back to the television and the dog.

She does not know what the software's internal acronyms mean. She has not read the product documentation. She could not care less about the mission statement. She may not have heard of the company at all, and if she has, she has not thought much about it. She does not know which team owns which part of the experience she is navigating, or what technical constraints shaped the decisions that produced it, or why the field she needs to fill in has a fifty-character minimum that nobody in the company that makes the software can explain.

She does not care about any of this. She should not have to. She just wants to complete her task and get on with her evening.

This woman — **distracted,** task-focused, entirely indifferent to the organizational structure of the software producer — is the entire point of the exercise. Every design decision in every digital product is ultimately made on her behalf, or should be. She is almost never in the room where those decisions are made. Someone has to be there for her.
That someone is the UX designer. That someone, in the meetings and workshops and sprint reviews and stakeholder presentations that make up the daily reality of product development, is the Dumbest in the Room.

The second person is my friend Albert.

His name is Albert. Al, to his friends. He is extremely helpful, unfailingly patient, and will do virtually anything you ask of him. He has read more than any human being who has ever lived. He can explain quantum physics, write a sonnet, translate legal documents, and produce a convincing imitation of almost any writing style you care to name. However, poor Al is not smart.

He also has six fingers on his left hand. You'll understand why in a moment.
Al has a superpower that makes everything else possible. He can stop time.
When you ask Al a question, he freezes the world — everything pauses, nobody ages, nothing moves — and he gets to work. He is infinitely patient in frozen time. He spends what would amount to years exploring every webpage, reading every book, working through every Wikipedia article, every forum post, every piece of text he can find. He doesn't sleep. He doesn't get bored. He reads everything.

Then, armed with all of this data, he starts looking for patterns. He doesn't understand your question — not really, not in the way you understand it — but he has read enough text to know what kinds of answers typically follow what kinds of questions. When he constructs a response, he's essentially asking himself: given everything I've read, what word is most likely to come next? And then again. And again. Until he has a sentence. Then a paragraph. Then an answer.

He unfreezes time and hands it to you. The answer is probably right. It is statistically derived from an enormous amount of human knowledge, applied at sufficient scale, which produces remarkably good approximations of correctness. But Al doesn't know if it's right. He can't know. He doesn't understand the question, and he doesn't understand the answer. He has found the most **likely** response based on everything humanity has written, and he's presented it to you in fluent, confident, well-structured prose.

Al has a few other qualities worth knowing about before you trust him with anything important. He believes everything he's read. Every conspiracy theory, every discredited study, every confidently wrong opinion published anywhere on the internet is in Al's training data, weighted alongside peer-reviewed research and established fact. He has no mechanism for distinguishing between them beyond the statistical signal of how often each appears. Popular misinformation, repeated frequently enough, can outweigh accurate information that appears rarely.

He is not creative, despite what the marketing materials suggest. Everything Al produces is a recombination of what humans have already made. He remixes, he synthesizes, he extrapolates — but he generates nothing that didn't exist in some form in his training data. The word "generative" in generative AI refers to the process of producing output, not to originating anything new. Al is the world's most sophisticated pastiche machine. That is genuinely useful. It is not the same as imagination.

He doesn't always know when he's wrong. This is the six-finger problem. Al has processed millions of images of human hands — but hands are complex, context-dependent things. Fingers overlap, curl, grip, hide behind objects. The statistical average of all those images doesn't resolve into a clear model of what a hand actually is. So Al produces hands that look approximately right from a distance and reveal their strangeness up close: too many fingers, joints at implausible angles, thumbs in unexpected locations. He's not malfunctioning. He's doing exactly what he does — finding the statistical center of a complex distribution — and the statistical center of "hand" turns out to be slightly wrong in ways that are immediately obvious to anyone who has spent their life in possession of one. Although this six-finger problem has largely been solved, it is an indicator of how AI works.

The same principle applies to facts. Al's confident wrongness about hands is the same confident wrongness that produces hallucinated legal citations and invented statistics. He doesn't know he's wrong because he has no model of wrongness. He has a model of likelihood. And likely is not the same as true.
Here is what Al is genuinely good at, and it is considerable.

He compresses. The world produces more information than any human can process, and AI can reduce that flood to something navigable. Ask him to summarize, synthesize, or identify the relevant parts of a large body of material, and he performs this task faster and more patiently than any human researcher. Sam Altman, OpenAI's CEO, has called this compression — the distillation of vast amounts of data into digestible form. It is a real and valuable capability.

He drafts. First drafts of almost anything — emails, reports, outlines, code — can be produced faster with AI's help than without it. The draft may need significant revision. It may be wrong in places. But it is a starting point, and starting points have genuine value.

He delegates the noise. The 80/20 principle suggests that roughly 20% of the decisions in any problem are the crucial ones, and the rest is information management. AI handles the information management, freeing the human to focus on the 20% that actually requires judgment. This is the complementarity argument we'll return to throughout this book — the idea that AI and humans are most valuable in combination, each handling what the other can't.

What AI cannot do is think. Not in any meaningful sense of the word. He has no awareness, no understanding, no goals, no curiosity, no sense of what matters. He is, despite the name we've given him, not intelligent. He is extraordinarily capable, which is a different thing entirely — and a thing that the industry has been at some pains to obscure, because "extraordinarily capable statistical pattern-matching system" doesn't generate the same venture capital interest as "artificial intelligence." The confusion between capability and intelligence is not a minor semantic quibble. It shapes how people trust these systems, how organizations deploy them, and how designers build interfaces around them. A system that is intelligent deserves a different kind of trust than a system that is capable. It warrants different oversight, different skepticism, and different design.

AI is very useful. AI is not intelligent. Both of these things are true, and holding them simultaneously is the beginning of using AI well.
The six fingers are a reminder to check his work.

What This Book Is About

Over thirty years I have watched the field evolve from "make sure the buttons are visible" to "deploy machine learning to predict and pre-empt every user action before the user has consciously formed it." That is an extraordinary leap. And the ethical and human framework for thinking about it has not kept up.

This book is my attempt to close that gap — or at least to describe it clearly enough that designers, technologists, product managers, and curious users can see it for what it is.

The title comes from two ideas that sound compatible but aren't.
DON'T MAKE ME THINK is Steve Krug's famous formulation from his 2000 book of the same name — the principle that good design should be so intuitive that users never have to stop and puzzle over what to do next. It remains correct. Unnecessary complexity is still the enemy of good design.

BUT DON'T THINK FOR ME is what happens when that principle is taken too far — when automation removes not just unnecessary friction but necessary agency, when the system is so confident in its model of you that it stops checking, when technology that was supposed to work for you starts working on you instead.
The space between those two instructions is where this book lives. It is large enough to contain thirty years of stories, several hundred dollars of personally incurred airline fees, at least one refrigerator with opinions about yogurt, and a German professor who gave free association a number.

Who This Book Is For

If you are a UX designer, this book is for the version of you that got into this work because you cared about people — and has spent subsequent years navigating the gap between that motivation and the organizational, commercial, and technical pressures that push against it. I hope it is useful, and honest, and occasionally a relief.

If you are not a UX designer but you use technology — which is to say, if you are a person alive in the twenty-first century — this book is for the version of you that has felt, at some point, managed rather than helped by a digital product. Second-guessed, herded, or quietly manipulated by something that was supposed to make your life easier. That feeling is not a personal failing. It is a design failure, and you were right to notice it.
Either way: welcome. The Dumbest in the Room is glad you're here.

Let's jump in.

Chapter 1: The Promise and the Problem

Bad UX design can be confusing. O'Hare Airport, Chicago

Great stuff, UX. But, like escorting an old lady across the road who didn't ask for it, the UX designer often assumes the user doesn't want to think; they want to be guided. However, the designers' assumptions can block and frustrate the user who does not want that guidance.

We'll look at several of these shortly — including a flight to Barcelona that cost me $150 in ways that were entirely avoidable.

Technology is great. Humanity is already addicted and completely dependent on it. There's a moment — and if you've owned a smartphone for more than six months, you know exactly the moment I mean — where technology does something so unexpectedly perfect that you actually look around the room to see if anyone else witnessed it.

Maybe your music app queued up the exact song you were about to search for. Maybe your navigation app rerouted you around a traffic jam you didn't even know existed yet, and you arrived at your destination two minutes early feeling rather pleased that something in your life was made easier. Maybe your email client drafted a reply so accurate you just hit send, went back to your coffee, and thought: *okay, we're living in the future and honestly, it's pretty great.*

That feeling? That's the promise.

Now let me tell you about my refrigerator.

I bought a smart refrigerator. I want to be clear that I did not need a smart refrigerator. My previous refrigerator was a perfectly competent appliance that kept things cold and asked nothing of me emotionally. But the new one had a screen. And an app. And it could, according to the box, "learn my household's consumption patterns and proactively suggest grocery needs."

I was supposed to bar-code everything I put in the fridge – it would then let me know when ordering groceries if I had too much or not enough of something. However, that requires discipline. And in a household of teenagers, that discipline went out the window almost immediately. So now I get notifications about a lack of milk, when two gallons had been bought but not bar-coded. Or when the kids put empty containers back in the fridge (why, I don't know) the fridge tells me that I have enough milk.

It also once sent me a push notification at 11:47pm to inform me that my yogurt was "approaching optimal consumption window."

That's not a smart refrigerator. That's an anxious one.

This is a book about the seam between those two experiences.

Between the music app that reads your mind and the refrigerator that pesters you about dairy products in the middle of the night. Between technology that makes you feel capable, even powerful — and technology that makes you feel managed, second-guessed, and quietly furious.

That space has a name. It's called user experience. And right now, in the age of artificial intelligence, it is both the most exciting and the most troubled it has ever been.

We Were Promised Frictionless

Back in 2000, Steve Krug's book called *Don't Make Me Think* was a manifesto. His argument was elegant: good design should be so intuitive that users never have to stop and puzzle over what to do next. No confusion, no hesitation, no hunting around for the button. Just flow.

It became one of the most influential design books ever written, and for good reason. Krug was right. Unnecessary complexity is the enemy of good design. Confusion is not a feature.

But here's the thing nobody fully anticipated in 2000: we would one day have technology capable of doing the *thinking for us entirely*. And that turns out to be a completely different problem.

"Don't make me think" and "think for me" sound like they're on the same side. They're not. One respects your intelligence. The other quietly replaces it.

The Thermostat That Knew Best

The Nest Learning Thermostat arrived in 2011 promising to "learn your schedule" and "program itself." For many people, it delivered. It quietly mapped their routines, adjusted temperatures before they needed adjusting, and saved energy without requiring a mechanical engineering degree to configure — which, if you ever owned a pre-smart thermostat, you'll appreciate is not a low bar.

But a recurring complaint in the reviews is telling. Users would come home to a house that was the wrong temperature. Not dramatically wrong — just slightly, subtly, *annoying* wrong. And when they tried to correct it manually, the Nest would sometimes interpret their override as new data and adjust its model accordingly, learning the wrong lesson entirely. The thermostat was too confident in its own intelligence. It had stopped checking in.

The users weren't cold. They were *dismissed.*

That distinction matters enormously in UX design, and we'll come back to it repeatedly throughout this book. There's a difference between a system failing and a system succeeding in a way that ignores you. The second one is much more corrosive to trust, because it suggests the system wasn't really paying attention to *you* at all. It was paying attention to a statistical model of you. Which is flattering in theory and maddening in practice.

The New Bargain

Every time you use a digital product today — a phone, an app, a website, a car, a refrigerator with opinions about your yogurt — you are entering into an implicit bargain.

The product says: *trust me. I know what you need. I've been designed to anticipate you, guide you, and smooth the rough edges off your experience. Just go with it.*

And most of the time, we do. Because most of the time it works well enough, and we're busy, and learning new interfaces is tedious, and honestly who has time to read the settings menu.

But the bargain has fine print.

The fine print says: *in exchange for this convenience, you will occasionally be herded, manipulated, second-guessed, profiled, and presented with the illusion of choice while the actual choices are made for you by an algorithm trained on the behavior of millions of other people who are not you.*

That's not a criticism of technology. It's a description of the core problem. And it's a problem that designers — the people who build these experiences — have a professional and, I'd argue, moral obligation to understand and take seriously.

Why This Book Exists

I've been a UX designer for thirty years. I've worked with banks, telecoms, healthcare companies, universities — organizations whose digital products are used by millions of people who largely have no choice but to use them. You don't shop around for your health insurance portal. You don't comparison-test your employer's HR system. You just use it, and you either feel respected by it or you feel like a rat in a maze.

In that time I've watched the field evolve from "make sure the buttons are visible" to "deploy machine learning to predict and pre-empt every user action before the user has consciously formed it." That is an astonishing leap in thirty years. And the ethical framework hasn't kept up.

We have more capability than we have wisdom about using it.

This book is not a manifesto against AI. I use AI tools every day and find several of them genuinely, embarrassingly useful. It's also not a nostalgic argument for making things harder. Friction for friction's sake is just bad design with a philosophy attached.

What this book *is*, is an argument for a specific kind of respect. The kind that says: yes, anticipate the user. Yes, reduce unnecessary effort. Yes, use the data you have to make the experience smarter. But never forget that there is a human being on the other end of this interaction, and that human being would quite like to remain in the driver's seat of their own experience.

Don't make me think.

But please, for the love of all that is usable — **don't think for me.**

What's Coming

The next nine chapters are going to take you through the full landscape of this tension. We'll look at where automation gets it gloriously right and catastrophically wrong. We'll spend some uncomfortable time in the world of Dark UX — design that is intentionally built against the user's interests, dressed up in friendly colors and rounded corners. We'll talk about trust, and what it actually takes to build it, and how quickly it can be destroyed. We'll look at what happens to humans — cognitively, psychologically — when technology does too much for too long. And we'll end with what good design actually looks like when it takes all of this seriously.

Along the way there will be anecdotes. Some funny, some less so. There will be examples from products you use every day.

Let's get into it.

Chapter 2: The Goldilocks Zone of Automation

In which we discover that "just right" is harder than it sounds, that an 1837's fairy tale has more to say about AI than most white papers, and that the correct amount of friction is not zero.

Let me tell you about two GPS experiences.

The first one: you're driving in an unfamiliar city, running late, slightly stressed. Your phone calmly says "turn left in 400 meters," and you turn left, and traffic is clear, and you arrive at your destination with three minutes to spare. The app has invisibly absorbed about forty variables — traffic density, road works, your current speed, three alternative routes — and distilled them into one simple instruction. It did the work so you didn't have to. You feel capable. You feel *helped.*

The second one: you're driving a route you know well — your own neighborhood, a road you've driven five hundred times. Your navigation app, which you left running out of habit, confidently announces that you should "take a slight left" onto a street you've known since 1987. You ignore it. It recalculates. It suggests the same street again, this time with slightly more urgency, as if you perhaps didn't hear it the first time. You ignore it again. It sulks briefly, then announces it has "found a faster route" — which is your original route, renamed.

Same technology. Same app. Two completely different experiences.

The difference isn't the algorithm. The difference is whether the automation was doing something useful *for you* or just doing something *to you*.

Too Cold, Too Hot, Just Right

In 1837, a woman named Eleanor Mure wrote a story about a little girl who breaks into a bear family's house, eats their food, sits in their chairs, and sleeps in their beds, choosing the option that's not too extreme in either direction. The story was refined over the following decades, the girl was renamed Goldilocks, and the underlying principle — that there's a zone between two bad extremes where everything works — became one of the most useful concepts in existence.

It applies, as it turns out, to AI automation with almost embarrassing precision.

Too little automation and users are overwhelmed. They drown in decisions, cognitive load piles up, and the technology that was supposed to help becomes a burden. Anyone who's ever tried to configure a professional audio interface, set up a home router, or file taxes without software assistance knows this feeling. It is not enabling. It is exhausting.

Too much automation and something more insidious happens. Users disengage. They stop understanding what's happening and why. They lose the ability to catch errors because they've stopped paying attention. They become passengers rather than drivers. And when the system eventually gets something wrong — which it will — they're not ready for it, because they've been systematically trained not to be.

The sweet spot in the middle has a proper name in UX and AI research. It's called **calibrated automation** — and it is significantly harder to design than either extreme, because it requires the system to be genuinely intelligent about *when* to help, *how much* to help, and crucially, *when to get out of the way*.

The Brilliant Intern Problem

Here's a useful mental model. Imagine you have a brilliant intern. This intern is extraordinarily fast, has read everything, makes connections you might miss, and can produce a first draft of almost anything in thirty seconds. Fantastic.

Now imagine this intern has no social awareness whatsoever. They answer questions you haven't asked yet. They complete your sentences out loud while you're still forming them. They have pre-emptively scheduled three meetings based on what they think you probably need. They've replied to two of your emails on your behalf. They have opinions about all of it.

At what point does "brilliant" become "unbearable"?

The answer is: roughly the moment they stop checking in.

A brilliant intern who asks "should I draft a reply to this, or would you prefer to handle it yourself?" is a gift. A brilliant intern who just... drafts the reply, sends it, and tells you afterward, is a liability. Not because they're wrong, necessarily. But because *you* are the one who's **accountable** for that email. You're the one whose name is on it. The decision was yours to make, and it was taken from you.

This is the fundamental tension in AI-augmented user experience, and it shows up everywhere from your email client to your car to your hospital's diagnostic software. The system's capability is not in question. The question is always: *whose decision is this, and does the design reflect that?*

The Seam

One of the concepts that stuck hardest — and one that I keep returning to in practice — is something called **the seam**— the handoff points between what the system does automatically and what the human is expected to do next.

The seam sounds boring. It is not boring. The seam is where most AI UX fails.

Think about Tesla's Autopilot. The system can handle highways, maintain lanes, follow traffic, and brake autonomously. Impressive. But early versions of the feature had a specific, nasty failure mode: they were so smooth, so competent on ordinary roads, that drivers disengaged mentally. They stopped monitoring. They started reading, eating, looking at their phones. And then, on the occasions when something unusual happened — a lane marking that confused the sensors, an unusual intersection — the car would politely ask the driver to take over.

With two seconds' notice.

Two seconds to go from "I was basically napping" to "I have full situational awareness and am making good driving decisions at 70 miles an hour."

That is a badly designed seam. Not because the automation was bad. Because the handoff was designed by people who thought very carefully about the driving part and not carefully enough about the *human* part.

A well-designed seam is visible. It tells you what the system is doing and why, right now. It lets you intervene cleanly, without fighting the system. It gives you a realistic path back into control. It treats the handoff not as an edge case but as an essential part of the experience.

The Helpful Shove

There is a particular kind of unhelpfulness that wears helpfulness as a disguise.

You know the type. The colleague who answers a question you didn't ask. The app that completes your sentence in a direction you weren't going. The navigation system that reroutes you onto a road you've been avoiding for reasons the algorithm has no way of knowing. Each of these is, in its own way, a shove — well-intentioned, firmly applied, and deeply annoying to the person being shoved.

I once had a colleague who was enormously proud of a website he'd designed. The navigation was hidden. To find the links, you had to explore — hover in the right places, discover the structure through a process of trial and intuition. He had designed it as an experience, a gentle game of discovery.

The bounce rate was extraordinary. Not in a good way.

He was genuinely surprised. The client was less surprised and considerably less gentle about expressing it. The website had been designed for the designer's idea of an engaged, curious user who wanted to play along — and had been encountered by actual users, who wanted to find the thing they came for and leave.

This is the original sin of a certain kind of UX design: the assumption that the user wants to be guided, shepherded, assisted, and occasionally delighted, when what they actually want is to accomplish something and get on with their day.

It is, as I have come to think of it, the problem of escorting an old lady across the road who didn't ask to cross.

The Barcelona Incident

Let me tell you about a flight to Barcelona that cost me $150 I did not intend to spend.

I was checking in online — my wife and I, straightforward enough. The page helpfully auto-filled our details from previous bookings, which is a genuinely useful feature for the repetitive fields: address, contact number, the kind of information that doesn't change between trips.

When I got to my wife's details, the autofill populated her last name with mine. Reasonable assumption on the system's part — many couples share a surname. My wife, however, travels on her passport under her maiden name, which is legally hers and the only name that matters at a border crossing.

I corrected her last name to her legal name and moved on.

What I did not notice — because it happened below the fold, silently, without any indication that it had occurred — was that the system had interpreted my correction as new information and helpfully updated my name to match hers.

The system was thinking for me. It had a theory about what I meant, and it acted on that theory without asking, without flagging, and without giving me any visible indication that it had done so.

I discovered this after I had completed the booking and paid. The correction cost $150 and a phone call that I would generously describe as character-building.

The automation was not malicious. It was, in its own logic, trying to help — maintaining consistency across the booking, applying a pattern it had learned from other users' behavior. But it was thinking for me in a context where the thinking was mine to do, on data where the consequences of being wrong were real and immediate, without showing its work or giving me the chance to check it.

This is the failure mode that the industry consistently underestimates: not the dramatic AI error, but the quiet, confident, invisible one. The assumption made below the fold. The correction applied without notification. The system so sure it knew what you meant that it didn't bother to ask.

I'm Not Finished Yet

There is a specific species of UI behavior that I find more irritating than almost anything else in digital design, and I encounter it approximately twice a week.

You are filling in a form. You reach a telephone number field. You type the first digit.

An error message appears. "PLEASE ENTER A VALID PHONE NUMBER."

You have typed one digit. You are aware, as a person who has had a telephone number for several decades, that one digit is not a complete telephone number. You were not under the impression that you had finished. You were, in fact, in the process of continuing to type when the interface decided to interrupt you with a correction for an error you had not yet made.

This is validation logic applied without any model of the user as a person in the middle of doing something. It is the digital equivalent of someone leaning over your shoulder while you're writing a sentence and saying "that's not a word" after the third letter.

The intention behind real-time validation is good: catch errors early, before the user submits and has to start over. The execution, in its most aggressive form, produces an experience that is irritating, demeaning, and in some cases actively confusing — particularly for users who are less confident with technology and interpret the error message as evidence that they are doing something fundamentally wrong, rather than simply not finished.

The fix is trivial. Validate on blur — when the user leaves the field — not on every keystroke. Wait until the person has finished before telling them they've made a mistake. This is a design decision that takes thirty seconds to make and produces a meaningfully better experience for every user who encounters it.

The Intern's Default

Some of the most consequential design decisions in the history of enterprise software were made by people who had no idea they were making design decisions at all.

I was working on a UX project for a Fortune 500 company — a data entry interface used by very senior managers who were, by the time I arrived, visibly frustrated with one particular field. The field had a fifty-character minimum. You could not submit the form without entering at least fifty characters in this field, regardless of whether fifty characters were appropriate for what you needed to say.

The frustration had reached the point where people were discussing building a separate application just to route around this one field. The workarounds being employed by senior executives to satisfy a fifty-character minimum — padding entries with spaces, adding meaningless qualifiers, typing the same phrase twice — were a small masterpiece of human adaptability in the face of arbitrary constraint.

I asked why the minimum existed.

Nobody knew.

I investigated. The field connected to a database. The database had a fifty-character minimum on that column. The minimum had been set during the database installation. The installation had been performed by an intern. The intern had set the minimum at fifty characters because that was what he had learned at school was good practice for keeping a database efficient — a principle that applied in certain technical contexts and had been applied here without any consideration of whether those contexts matched.

An intern, following a rule learned in school, applied without thought to a context he didn't fully understand, had been frustrating the working lives of senior managers at a major corporation for several years.

Nobody had questioned it because the field had a minimum, and fields have minimums for reasons, and querying every technical decision in an enterprise system is not how organizations function. The assumption was that someone had decided this. The reality was that a default had decided it, and the default had been set by someone who was thinking about database efficiency and not thinking about users at all.

The fix took an afternoon. The field minimum was removed. The workarounds stopped. The separate application was never built.

The lesson — that defaults are policy decisions whether or not anyone treats them as such, and that the cost of an unexamined default compounds silently over years — is one that the industry relearns, expensively, on a regular basis.

The Point of All of This

The thread connecting the hidden navigation, the Barcelona booking, the impatient form validator, and the intern's database minimum is not that technology fails. Technology fails constantly and we manage.

The thread is that each of these failures came from a system — or a designer, or an intern — that was thinking for the user without thinking ABOUT the user. Making assumptions about what the user meant, what the user needed, what the user should be allowed to do, without asking, without flagging, without giving the user the information or the opportunity to decide for themselves.

Good UX does not assume. It informs, suggests, and gets out of the way.

It does not complete your sentence before you've finished it, correct your name to someone else's, or tell you that three is not a valid number while your finger is still on the key.

It does not escort you across the road without asking where you were going.

The Slider That Nobody Moves

One of the more interesting ideas in AI UX research is **adjustable autonomy** — the idea that users should be able to tune how proactive a system is, dialing it up when they trust it and have low-stakes decisions, dialing it back when they're in unfamiliar territory or the consequences of error are significant.

In theory, this is excellent design. In practice, there's a catch.

Nobody moves the slider. It is very common that elaborate profiles can be created that would be immensely useful to the user, to give them a bespoke experience. These are often just ignored. Nobody reads the manuals, people just jump in.

Default settings are essentially permanent settings for the overwhelming majority of users. Research consistently shows that people accept defaults at staggering rates, regardless of whether those defaults serve them well. Cookie consent, privacy settings, notification preferences, subscription renewals — the default wins, every time, for most people.

Which means the most important design decision in adjustable autonomy is not the existence of the slider. It's **where you put the default**.

If the default is "high automation," most users will run in high automation mode forever, including in situations where it's not appropriate. If the default is "conservative, suggestion-only," most users will benefit from that caution even when they didn't consciously choose it.

There's a lot of money in defaulting to high automation. Products look more impressive in demos. Engagement metrics go up. Users feel the magic. The business case for conservative defaults is harder to make in a quarterly review.

This is one of the reasons that designing honestly in the Goldilocks zone requires more than good intentions. It requires someone in the room with enough standing to say: we're making this too autonomous, and that serves us, not the user. Those conversations are harder than they sound.

What "Just Right" Actually Looks Like

Let me give you a concrete example, because "calibrated automation" is one of those phrases that sounds sensible and means nothing until you see it in practice.

Google Docs has a feature called Smart Compose. As you type, it offers greyed-out suggestions for how to finish your sentence. You can accept a suggestion by pressing Tab, or just keep typing and it disappears.

This is good Goldilocks design, and here's why.

First, it's **transparent**. You can see exactly what it's suggesting before you accept it. There's no mystery, no "the system did something and now I have to figure out what."

Second, it's **opt-in by action**. You have to press a key to accept. It does not complete your sentence and then look at you expectantly. You remain in charge of every word.

Third, it **fails gracefully**. When Smart Compose gets it wrong — which it does, regularly — nothing bad happens. You just keep typing. There's no damage to undo, no confused state to unpick.

Fourth, and most underappreciated: it **makes you faster without making you dependent**. If Smart Compose disappeared tomorrow, you'd still be able to write. You haven't offloaded the skill. You've just borrowed a little help.

Compare this to an email client that automatically categorizes, prioritizes, summarizes, pre-drafts responses, and schedules follow-ups — all without asking. That system might handle each of those tasks more "efficiently." But taken together, what it's doing is systematically removing you from your own inbox. At what point is it your email?

The One-Line Test

From my studies, I came across a discipline I've since used with teams called the **AI contract** — a single sentence that forces clarity about what exactly an AI feature is doing.

The template goes: *Given [inputs], the system produces [outputs] to help the user [action], under [constraints].*

It sounds simple. It is remarkably hard to write honestly.

"Given your emails, the system produces suggested replies to help you respond faster" — clean, specific, testable.

"Given your behavior, the system produces a personalized feed to help you discover content you'll enjoy" — sounds reasonable until you notice that "content you'll enjoy" and "content that maximizes your time on platform" are not the same thing, and the contract doesn't say which one is actually being optimized.

The AI contract is a Goldilocks detector. If you can't write it cleanly, the feature probably isn't calibrated well. Either it's doing too little (no clear output or action) or too much (the action belongs to the system, not the user), or the constraints are missing — which is usually where the mischief lives.

Where the Zone Actually Lives

So where, practically, is the Goldilocks zone?

After thirty years of designing for humans and several years specifically in AI-augmented contexts, my working answer is this: **the zone is wherever the user remains capable**.

Suggest before you act. Show your reasoning. Make correction cheap. Make undo real. And every time you're tempted to remove a decision from the user because it's more efficient, ask yourself one question first: *when this goes wrong — not if — will the user be ready?*

If the answer is no, you've overshot the zone. Don't make the user think about the technology. Let them think for themselves.

The Goldilocks Zone is not just about how much automation to apply. It's about whether the automation respects the person it's supposed to be serving.

Chapter 3: When the Machine Gets It Wrong

In which we learn that confidence and competence are not the same thing, that a single bad prediction can undo months of goodwill, and that the most dangerous words in technology are "based on your preferences."

In the summer of 2016, a man named Robert Julian-Borchak Williams was arrested at his home in Detroit. He was handcuffed in front of his wife and daughters, taken to a detention center, and held overnight before being interrogated. The charge was felony theft.

The evidence against him was a match from a facial recognition system.
The system was wrong. Williams, who is Black, had been incorrectly matched to surveillance footage of a shoplifter. When investigators showed him the still image during questioning — a grainy photograph of a man who looked, by any reasonable assessment, nothing like him — Williams held it up next to his own face and said: *"I hope you don't think all Black men look alike."*

The charges were eventually dropped. Williams filed a lawsuit. The story became one of the most cited examples in the ongoing legal and ethical debate around automated decision-making.

This is an extreme case. Most AI errors don't result in wrongful arrests. Most of them are smaller, quieter, and less dramatic. A medical scan misread. A loan application incorrectly scored. A job candidate filtered out by a résumé screener that turned out to be biased against graduates of all-women's colleges. A credit card flagged for fraud on the day you're trying to buy a plane ticket home for your mother's funeral.

What these cases share is structure. In every one of them, a system produced a confident output, that output shaped a significant decision, and somewhere in the chain, the human who should have been checking was either not looking, not empowered to push back, or not given the tools to question what the machine had decided.

That is a UX problem. Fluent Is Not Factual

Here's something worth understanding about how modern AI language systems work, because it explains a failure mode that is genuinely new in the history of technology.
Large language models — the kind that power chatbots, writing assistants, summarization tools, and a growing number of enterprise applications — are, at their core, extraordinarily sophisticated prediction engines. They are trained on vast amounts of data and have become exceptionally good at predicting what words should follow other words. The result is output that reads as fluent, confident, and well-structured.

What it does not mean its *correct.*

This is such an important distinction that it's worth saying twice. Fluency and accuracy are independent properties. A system can produce beautifully written, grammatically impeccable, completely confident nonsense. It can cite sources that don't exist, quote statistics that were never gathered, and summarize documents in ways that subtly but materially misrepresent what those documents say — all in the same assured tone it uses when it's perfectly right.

The technical term for this is hallucination, which is a slightly whimsical name for something that has caused some genuinely un-whimsical problems.
In 2023, a New York lawyer named Steven Schwartz submitted a legal brief that cited several court cases in support of his client's position. The opposing counsel, unable to locate these cases, asked for copies. It turned out they did not exist. The lawyer had used ChatGPT to help research the brief and had not verified the citations. The cases were entirely fabricated, complete with invented judges, invented courts, and invented rulings, all written in the measured, authoritative language of actual legal precedent.

The judge was not delighted. Schwartz and his colleague were sanctioned. The story ran in every major newspaper for a week.
The lesson here is not "don't use AI tools." The lesson is: the interface gave no indication that the confident-sounding output might be invented. There was no uncertainty cue. No "I'm not entirely sure about this one." No flag on the citations saying "you may want to verify these exist before submitting them to a federal court." Just smooth, professional-looking text, formatted like something you could trust.

That's a design failure. The model was doing what models do. The interface failed to communicate the model's limitations.

The Confidence Problem

Traditional software, when it was wrong, tended to be wrong in obvious ways. The calculation returned an error. The field highlighted in red. The page failed to load. These were annoying, but they were legible. You knew something had gone wrong. AI systems introduce a genuinely new failure mode: the system is wrong, but nothing looks wrong. The output is presented with the same visual authority as correct output. Same font, same formatting, same little checkmark or completion animation. The interface has no way to express doubt, because it wasn't designed to.

I once noted this down during a course and haven't been able to improve on it since: *"The system can be confidently wrong, plausibly worded, and still useless."* That's not just a technical observation. It's a design brief. It's telling you that every AI-powered interface needs to answer the question: *how does a user know when to trust this?*

The answer, almost always, is evidence. Not explanation — evidence. Users don't need a lecture on how the model works. They need to see the sources. They need to see the confidence range. They need to see what the system is basing its output on, so they can make a judgment about whether that basis is solid.

A medical AI that says "this scan shows no significant anomalies" is less useful than one that says "this scan shows no significant anomalies — compared against 340,000 reference scans, similarity index 94%, three regions flagged for human review." The second version is not more complicated. It's more honest. And it keeps the radiologist in the loop rather than gently elbowing them out.

The Trust Bank Account

Think of user trust as a bank account. Every time the system works well — anticipates correctly, saves time, reduces effort — a small deposit goes in. Every time it fails visibly — wrong recommendation, missed context, embarrassing autocomplete — a withdrawal happens.

The account starts at zero, or close to it. Users don't trust new systems; they extend provisional credit, which gets confirmed or withdrawn based on experience.

Here's the asymmetry that matters for designers: withdrawals are larger than deposits.

This is well-documented in behavioral psychology — losses feel roughly twice as significant as equivalent gains — but it's particularly acute for AI systems, and for a specific reason. When a human assistant makes a mistake, you recalibrate your expectations of that person. When an *algorithm* makes a mistake, something more fundamental shifts: you start to question the entire premise.

Researchers call this **algorithm aversion**. It was documented in a 2015 study by Dietvorst, Simmons, and Massey, and the finding is stark: people who see an algorithm make a mistake will subsequently prefer a less accurate human to a more accurate algorithm, even when shown data proving the algorithm outperforms. One visible failure is enough to trigger a disproportionate, lasting loss of trust — even when the system is still better than the alternative.

This should terrify anyone shipping AI features.

It means that the one time your product recommendation engine suggests something bafflingly wrong — the time it suggests a book about grief to someone who just bought a birthday card, or recommends a restaurant to someone who reviewed it one star six months ago — is not just an amusing edge case. It's a trust withdrawal that will shape how that user interacts with the feature for months.

The implication for design: make errors less visible where they're low-stakes, and build recovery into every high-stakes flow. When the system gets it wrong, it should be easy to say so, easy to correct, and ideally, the correction should demonstrably improve future behavior. A system that learns from your corrections is a system that earns trust back. A system that repeats the same error is a system you stop using.

The Embarrassment Radius

There's another dimension to AI errors that doesn't get discussed enough: embarrassment.

Not just errors that cost you money or time — errors that cost you dignity. Autocorrect has a rich and well-documented history of replacing perfectly reasonable words with things that no adult would write in a professional email. Predictive text has sent messages to the wrong people, with the wrong sentiment, at the worst possible moment. AI-generated email summaries have missed the single most important sentence in the thread — the one that changed everything — and presented a confident summary that left the reader completely blindsided in the meeting they were supposedly prepared for.

These errors have an outsized psychological impact because they involve the user's social presentation. You didn't just get wrong information — you *looked* wrong, to someone whose opinion of you matters. You sent the email. You walked into the meeting. You trusted the summary.

This is why the stakes in AI UX are higher than in traditional software design. When a button doesn't work, you're frustrated. When an AI system confidently produces something wrong and you act on it — in front of other people, in a consequential context — the damage is harder to undo. Trust isn't just lost in the product. Trust is lost in the person who used it.

That said, a teacher friend of mine would write nasty emails about some of his pupils to their parents telling them what he really thought of their brats, then used AI to make it "acceptable". It was easier to write, and kind of cathartic.

Silent Omissions: The Errors Nobody Catches

The most dangerous AI errors are not the ones that are obviously wrong. They're the ones that are *mostly right*. This is known as silent omissions — cases where the model ignores an important constraint, simply doesn't mention a critical caveat, or produces an output that is technically accurate but missing the one thing that would have changed the decision.

Imagine an AI-powered contract review tool. It reads a 40-page vendor agreement and surfaces the key terms, unusual clauses, and potential risks. It does this in thirty seconds, which would have taken a junior lawyer two hours. It's impressive. It's useful.

And it missed the clause on page 31 — the one about automatic renewal with a six-month written notice requirement — because that clause used non-standard language that fell outside its training distribution.

You signed the contract. You're now locked in for another year.

The tool wasn't wrong about anything it said. It just didn't say everything. And because the output looked complete — structured, professional, comprehensive — you had no reason to doubt it.

This failure mode is particularly treacherous because it requires users to know what they don't know. To catch a silent omission, you'd have to suspect something was missing, then go and find it yourself. But you used the tool *in order to **not** have to read the whole document*. That's the whole point.

The design response to silent omissions is not to make the AI more comprehensive — that's the model team's problem, not the design team's. The design response is to be explicit about scope and to build in exit ramps: moments in the flow where the system reminds you of its limitations and gives you a clear path to human review when the stakes are high enough to warrant it.

"This summary covers the main commercial terms. For regulatory compliance review or unusual jurisdictions, consider legal review before signing." That's a sentence. It costs nothing. It might save a client relationship.

The Recovery Problem

When things go wrong — and they will — how a system recovers matters as much as the error itself.

Bad recovery looks like this: the system gave you wrong information, you discovered it too late, there's no easy way to flag the error, no clear path to correct your downstream decisions, and when you try to undo, you find the undo button doesn't actually work for this type of action.
This experience doesn't just make you distrust the feature. It makes you distrust the *company*.

Good recovery is almost a form of grace. The system acknowledges the error. It makes correction simple. It doesn't punish you for pushing back with extra steps or lost progress. And ideally, it treats your correction as useful information — something that makes the system slightly better for next time.

The best way to design recovery is to make it a first-class design requirement from the start, not an afterthought bolted on after launch. This means asking, for every significant AI-driven action: *what happens when this is wrong?* Not "if." *When.* Build the answer to that question into the interface before you ship. The cost of doing it then is far lower than the cost of doing it after your users have already made the withdrawal from the trust bank account.

A Brief Note on Apologies

Sorry, one last thing.

There is a category of AI error recovery that has become almost parodically bad, and it is the AI apology.

"I'm sorry, I seem to have made an error." "I apologize for any confusion this may have caused." "I'm sorry you experienced that."

These are not apologies. They are apology-shaped objects. They have the grammatical structure of remorse but none of the substance, because an AI system cannot be remorseful — and more to the point, the apology does nothing to fix the problem or prevent it from happening again.

Worse, they create a kind of uncanny valley of accountability. The system sounds like it's taking responsibility without actually doing anything about it. Which is, if you think about it, the worst possible combination: the optics of contrition without the mechanics of repair.

If your system makes an error, the interface should help fix it, and hopefully even learn from it. Not apologize for it. There's a difference.

Interlude: Rembrandt's Beard

I should tell you that I was a professional illustrator once. When I started working on a book that required a scene involving Rembrandt — the seventeenth-century Dutch master, in his studio, talking to a young woman — I needed an illustration. I did not have time to draw one. I decided to give AI image generation a try.

My prompt was: Rembrandt talking to a girl about a portrait he is painting with a lot of ultramarine.

> *A side note on ultramarine: it was ground from lapis lazuli, extraordinarily expensive, and so symbolically loaded with ideas of virtue and divinity that it was reserved almost exclusively for the robes of the Virgin Mary. If you are ever standing in front of a seventeenth-century painting and wondering why everyone's clothes are so drab except for the one person in brilliant blue, now you know.*

The images came back. I fell off my chair.
They were extraordinary. Rich, atmospheric, technically convincing. The light was right. The studio felt right. The paintbrushes and pots and the smoky, candlelit atmosphere were all exactly what I'd been imagining. In one image, Rembrandt appeared to be telling a whimsical story, the young woman listening with visible amusement. In another, he was teaching, she was attentive. There were details I hadn't asked for — a figurine on a shelf, drips of ultramarine on the easel — that suggested genuine visual intelligence at work.

I was impressed. I was genuinely, unexpectedly, delightedly impressed.
And then I looked more carefully at the man in the paintings.

The Problem with Rembrandt

Rembrandt van Rijn painted more self-portraits than almost any artist in history. We know exactly what he looked like: broad face, bulbous nose, slightly pained expression, the general appearance of a man who has been asked to sit still for too long and is not enjoying it. He did not have a beard for most of his career. He was not, by the conventional standards of his time or ours, conventionally handsome.

The man in my AI-generated images was tall, dark, distinguished, and bearded. He looked like a romantic novel's idea of a Dutch Golden Age artist. He looked like what you would cast in the film if you were trying to make Rembrandt more palatable to a contemporary audience. He looked, in short, like everything Rembrandt was not.

This is the confidence problem made visual. The system had produced something that looked completely convincing, was beautifully rendered, and was wrong about the one specific thing I had actually needed it to get **right**. If I had been writing an editorial piece about Rembrandt and had published that image, there would have been letters.

I refined the prompt. Rembrandt not wearing a hat or a beard, talking to a woman about a portrait he is painting with a lot of ultramarine.

I fell off my chair again — this time for different reasons.
The images were, if anything, more beautiful. The candlelight was exquisite. There was what appeared to be a mouse in the bottom left corner of one of them, a detail so perfectly period-appropriate that I chose to believe it was intentional. I loved the first one so much I printed it and framed it. The man in it still had a beard.

The Escalating Beard

I tried again. I added the full name. Rembrandt van Rijn, not wearing a hat or a beard. The beard got longer.

One of the new images featured a man who had apparently abandoned painting to pursue a career as a pirate. He was wearing a hat of such elaborate plumage that it suggested either a very good costume department or an AI that had decided hats were non-negotiable regardless of instructions. In the background, clearly visible on the easel, was what appeared to be a black-and-white photograph — a medium that would not be invented for another two hundred years, but which the system had apparently decided was era-appropriate.

I gave it one final attempt. The real Rembrandt van Rijn, not wearing a hat and clean-shaven, talking to a woman about a portrait he is painting with a lot of ultramarine.

The system replied, with the measured dignity of someone declining an unreasonable request, that it was unable to generate images of specific historical figures. It offered instead to produce "*a fictional scene inspired by the era in which Rembrandt lived, featuring an unnamed artist in 17th-century attire.*"

The unnamed artist, when he arrived, was extremely handsome, bearded, and wearing a hat.

What Ten Minutes With DALL-E Actually Taught Me

I want to be clear that I found this entire experience genuinely enjoyable. I am not telling this story as a complaint. The images were beautiful. The imagination at work — whatever we want to call the process that generates these things — was evident and impressive. I printed one and put it on my wall, which is a level of engagement I have not extended to most software in my career.

But the experience was a precise and rather elegant demonstration of everything this book has been arguing about in more serious contexts.

The system was confident. It produced outputs that looked exactly right. The wrongness was not visible in the rendering — it was only visible to someone who already knew what Rembrandt looked like, which is to say, someone who already had the knowledge the system was supposed to be providing. A user who didn't know would have accepted the handsome bearded stranger without question.

The corrections did not correct. Each refined prompt produced a new image that addressed some of my concerns and introduced new ones. The beard, specifically, became a kind of running argument between me and the system — me insisting it be removed, the system insisting, with each iteration, on restoring it in a slightly more elaborate form. By the final version it had achieved a magnificence that suggested the system had developed a personal attachment to it.

The system eventually refused the task rather than getting it right. When pushed to produce the actual Rembrandt, it declined on policy grounds and offered me a fictional substitute — which was, in its way, an honest acknowledgment of its own limitations. But it was an acknowledgment that came only after several rounds of confident wrongness, not before.

And the output, for all its limitations, was useful. Not as a finished illustration — I still don't have an accurate portrait of Rembrandt for my book — but as raw material. As inspiration. As a demonstration of the visual territory I was working in, rendered in minutes rather than hours. The heavy lifting, as I noted at the time, had been done. The finishing was still mine to do.

This is, as it turns out, a reasonable description of what AI is currently good for. Not the last mile. Not the judgment call. Not the thing that requires knowing what Rembrandt actually looked like, or caring about the difference.

The heavy lifting. The inspiration. The starting point that gets you closer to the thing you're actually trying to make.

Everything after that is still a human job. The beard notwithstanding.

Chapter 4: The Disappearing User

In which we discover that knowing someone very well and understanding them are completely different things, that a system optimized for your past is a trap for your future, and that the most sophisticated user profile ever built might have you completely wrong.

There is a version of you that lives inside every major digital platform you use.

It has your name, your age, your location. It knows what you've bought, what you've searched for, what you've clicked on and — perhaps more revealingly — what you've hovered over without clicking. It knows what time you check your phone in the morning and how long you read before you give up. It has inferred your income bracket, your political leanings, your relationship status, your health anxieties, and — with unsettling accuracy — your emotional state at various points during the day.

This version of you is updated constantly. It is, in some measurable sense, the most detailed portrait of your behavior that has ever existed in human history.

It is also not really you.

It is a statistical model of your past behavior, projected forward as a prediction of your future behavior. It has no idea what you're trying to become. It doesn't know about the resolution you made last January, the book you're trying to finish, the job you're applying for, the person you'd like to be in five years.

It knows what you *did*. It has no theory of what you *want*. Most LLM's are not even recently trained, but who's world view is month's if not a year old.

This distinction — between the user the system has modeled and the actual human using it — is one of the most underexplored problems in UX design. And as personalization engines become more sophisticated, more pervasive, and more consequential, it is becoming urgent.

The Recommendation Engine That Knew Too Much

Netflix, in its earlier years, ran a famous competition. They offered a million dollars to any team that could improve the accuracy of their recommendation algorithm by 10%. It ran for three years, attracted over 40,000 teams, and produced a genuinely impressive result: the winning algorithm was significantly better at predicting what users would rate highly.

There was just one problem. Netflix never fully deployed it.

The reason, which a Netflix engineer later discussed publicly, was illuminating: the improved algorithm was good at predicting ratings but not particularly good at predicting *what people would actually watch*. It turned out that what you rate highly and what you'll actually press play on at 10pm on a Tuesday are different things, driven by different needs — mood, energy level, who you're watching with, how much cognitive effort you have left after a long day. Think Sofia on the sofa, in Chapter one.

The algorithm had gotten better at modeling a version of you — the version with considered opinions about film — and missed the actual human slumped on the sofa wanting something that doesn't require subtitles.

This is a relatively benign example. The recommendation engine learned something useful from it, and Netflix built towards something more nuanced. But it illustrates a structural problem that scales badly as personalization gets more aggressive: **the more a system optimizes for a model of you, the less it has to do with you.**

You Are Not Your Clickstream

There is an industry term — "the user" — that sounds specific but is doing a lot of concealment.

"The user" in most product analytics means: the aggregate behavioral profile constructed from logged interactions. What that profile captures is clicks, scrolls, purchases, time-on-page, search queries, and similar signals. What it does not capture is intention, context, regret, or growth.

Consider what your browsing history actually represents. You searched for symptoms of a condition you were worried about — now you're being shown health anxiety content. You bought a gift for someone else — now you're being shown more of that category as if it's your own taste. You clicked on an outrage-inducing headline because you couldn't believe it was real — now the algorithm has logged a preference for outrage-inducing content and is busy finding you more.

Each of these signals is real. Each of them is a misread.

The system isn't stupid. It's doing exactly what it was designed to do: find patterns in behavior and use them to predict future behavior. The problem is that behavior is a noisy, context-dependent, often unrepresentative signal for what a person actually wants or needs. And when you build an entire personalized experience on top of that noisy signal, you get something that feels like it knows you while actually knowing only a funhouse mirror version of you.

The worst version of this is what happens when the model is confident and wrong in a direction that reinforces itself.

The Filter Bubble (And Why It's Worse Than You Think)

In 2011, activist and writer Eli Pariser coined the term "filter bubble" to describe the phenomenon of personalization algorithms creating self-reinforcing information environments. You see content that matches your existing views, which confirms those views, which makes similar content more likely to appear, which further confirms those views.

No one designed this outcome. No one sat in a product meeting and said "let's make our users less informed and more tribal." It emerged from the optimization target. The target was engagement. The side effect was epistemic narrowing. And because the side effect was invisible in the dashboard — no column for "worldview diversity score" — it compounded unchecked for years.

This is the "artifacts have politics" problem that Langdon Winner already wrote about in 1980. The interface is not neutral. The algorithm is not neutral. Every design choice embeds a value, whether you intended it to or not. The question is not whether your design has values — it does — but whether you've thought about what they are.

The Personalization Paradox

Here is the thing about personalization that nobody in the product meeting wants to say out loud: **a perfectly personalized experience is a closed one.**

If a system perfectly predicts and serves everything you already like, you will never encounter anything that surprises you, challenges you, or expands what you know. You will be in a loop of your own preferences, served back to you with increasing precision, forever.

This is not a hypothetical. Spotify users who rely exclusively on algorithmically generated playlists report, at measurably higher rates, a sensation music listeners have taken to calling "Spotify fatigue" — a feeling that all the music sounds the same, that discovery has somehow stopped, that the catalog feels simultaneously vast and claustrophobic.

The algorithm didn't run out of music. It ran out of *your* music — the you it had modeled — and started repeating itself, because that model had been fully exploited.

The solution is not less personalization. It's personalization that's honest about its own limitations and deliberately builds in what designers sometimes call **serendipity engineering** — the intentional introduction of things the user hasn't asked for but might genuinely value. The good record shop that puts an album in the "staff picks" section you'd never have found yourself. The library that includes a browsing section. The algorithm that knows when to stop being an algorithm.

This is harder to build than it sounds, because serendipity doesn't optimize cleanly. You can't A/B test for the thing the user didn't know they wanted. It requires designers to make value judgments about what a good experience is beyond "high engagement" — and those judgments are uncomfortable because they involve deciding you know something about users' interests that the users haven't told you.

Which brings us to a harder question.

Who Is This Experience For?

Cast your mind back to any major platform's "For You" feed. The name is instructive. *For you.* Designed around your preferences. Built to serve your interests.

Now ask: whose interests is it actually serving?

The honest answer, in most cases, is: primarily the platform's. The "For You" feed is optimized for time-on-platform, ad impressions, and return visits. Your preferences are the *input* to that optimization, not the beneficiary of it. You are, in the terminology of tech criticism, not the customer. **You are the product.**

This is not a conspiracy. It's a business model. And business models are not inherently evil. But when the business model's optimization target diverges significantly from the user's actual interests — and in attention-economy products, they diverge constantly — the personalization that looks like a service is quietly functioning as a mechanism of capture.

The UX of this capture is sophisticated and largely invisible. The feed is smooth. The recommendations are often good enough. The frictions that might prompt you to stop and reflect — ads that feel like ads, prompts that remind you how long you've been scrolling, content that actively disagrees with your worldview — are systematically removed or minimized. The experience is optimized to keep you in it, not to serve you well while you're there.

There's a meaningful difference between those two things. Good UX design has always known this. The best designers I've worked with have an almost allergic reaction to dark patterns precisely because they represent a betrayal of the basic contract: *I am designing this experience for the person using it.*

When personalization serves the platform rather than the person, it's not really personalization. It's targeting. And targeting, dressed up in the language of care and customization, is one of the more insidious designs of the last twenty years.

The Ghost in the Profile

There's a specific experience that most heavy users of personalized systems eventually have, and it tends to produce a particular kind of unease.

You notice the system has you wrong in a way you can't correct.

Not wrong in a trivial way — wrong about something that matters. Maybe it's consistently showing you content that reflects a version of your political views you've moved away from. Maybe it's treating you as a customer in a demographic that doesn't reflect who you actually are. Maybe it's a health app that keeps pushing content about a condition you were researching for a family member two years ago and has decided is your own primary health concern.

You try to signal the correction. You skip the content. You mark it as "not interested." You update your preferences in the settings menu, which nobody ever uses because it's buried under four sub-menus and has thirty-seven options that require a graduate degree in information architecture to interpret.

And the system, unmoved, continues to show you content for the person it thinks you are.

This is a ghost. A behavioral ghost of a past version of you, still haunting your feed, still shaping your experience, immune to revision because the model has decided it knows you better than you do.

The UX failure here is not the model's inaccuracy. Models are imperfect. The failure is the **absence of legible user control** — a real, usable, effective way for users to say *this is wrong, please update* and have that update actually propagate meaningfully through the system.

Most "preference" controls are theater. They exist to satisfy regulators and to give users a feeling of agency without substantially altering the algorithmic output. Real user control would be expensive — it would require the system to respect inputs that contradict its model's confidence — and so it doesn't really exist.

Designing for the User Who Will Exist Tomorrow

Here is the design challenge that all of this leads to, and it is genuinely hard.

Most UX design is built around understanding who the user is now and serving that person efficiently. User research, personas, behavioral analytics — all of these are tools for modeling the current user.

But people change. They grow, they shift priorities, they go through experiences that reconfigure what they want from their technology. A person in grief needs different things from a music app than they did six months ago. A person returning to work after caring for an elderly parent has different energy and attention than they did before. A teenager turning into an adult is, in the most literal sense, not the same user they were.

A system that only knows your past has no model for your becoming. And as these systems become more embedded in daily life — not just recommending films but mediating job searches, financial decisions, health choices, and social connections — the cost of being locked into a model of your past self rises sharply.

The most honest thing a personalization system can do is hold its own model lightly. To treat its profile of you as a working hypothesis rather than a settled fact. To build in mechanisms for users to actively update and correct it. To occasionally offer something unexpected — not because it's predicted to work, but because growth requires it.

This is not currently how most of these systems are built. They are built to be confident. Confidence performs better in demos. Confidence feels like intelligence.

But in the long run, a system that remains genuinely curious about who you are today, rather than certain about who you were yesterday, is a system that earns something rarer than engagement.

It earns actual trust.

Chapter 5: Dark UX: Designed Against You

In which we discover that not all bad experiences are accidents, that some of the world's most talented designers are using their powers for evil, and that the "close" button is not always where it appears to be.

Everything we've discussed so far has been about failure.

Systems that got it wrong by accident. Algorithms that modeled you badly without meaning to. Automation that overstepped because nobody thought carefully enough about the handoff. Interfaces that buried your preferences under seven sub-menus not out of malice but out of organizational entropy and competing priorities and the fact that the product manager who owned that screen left the company in March.

This chapter is different.

This chapter is about the failures that aren't failures at all. The experiences that feel broken but are, in fact, working exactly as designed. The buttons in the wrong place on purpose. The cancellation flow that takes eleven steps because every additional step is a percentage point of retained revenue. The notification that arrives at 11pm not because the algorithm miscalculated but because 11pm is when you're most emotionally susceptible and someone, somewhere, ran the numbers on that.

These are not bugs. They are features. And the discipline of designing them has a name.

It's called Dark UX. And it is, depending on your perspective, either the most cynical application of design skill in the history of the profession, or just good business.

Spoiler: it's the first one.

A Brief and Uncomfortable History

The term "dark patterns" was coined in 2010 by UX designer Harry Brignull, who was so troubled by what he was seeing in the wild that he built a website to catalog them. The original taxonomy included classics like the Roach Motel ("easy to get in, impossible to get out"), Confirm-shaming ("No thanks, I don't want to save money"), Misdirection, Hidden Costs, and the Trick Question — the pre-ticked checkbox, usually buried in a wall of terms and conditions, that signs you up for a newsletter, a trial, a data-sharing agreement, or occasionally all three.

These patterns were not new in 2010. Tabloid newspaper subscription cards had been using them for decades. Mail-order companies had been hiding auto-renewal clauses in fine print since the invention of fine print. But the web gave dark patterns scale, precision, and — crucially — testability. You could now run an A/B test on exactly how much you could obscure the unsubscribe button before your churn rate started to drop. You could iterate toward maximum user confusion with the same rigorous methodology you'd use to improve a checkout flow.

Which is exactly what happened.

The Roach Motel

The Roach Motel is perhaps the purest dark pattern: a service that is trivially easy to join and structurally nightmarish to leave.

You know this experience. You signed up for a free trial with one click. Cancellation, you discover fourteen days later, requires logging in, navigating to account settings, finding a "Manage Subscription" link that is rendered in a slightly lighter grey than everything around it, clicking through to a retention flow that asks why you're leaving, offers you a discount, expresses concern for your wellbeing, suggests you might want to pause instead, asks if you're sure, asks again, and finally, on what feels like the moral equivalent of a DMV queue, presents a confirmation button that says "Cancel Anyway" — phrased specifically to make you feel like you're doing something slightly unreasonable.

Amazon Prime, at various points in its history, has required up to six clicks and several pages of retention messaging to cancel. The FTC eventually sued them over it. In 2023 they were ordered to simplify the process. The fact that it required a federal lawsuit to make a cancel button easier to find tells you everything you need to know about the incentive structure at work.

The Roach Motel works because inertia is a powerful force in human behavior. Every additional step in a cancellation flow is a moment where the user might give up, get distracted, or be talked out of it. Designed inertia is not neutral design. It is friction deployed as a retention strategy, and it works well enough that very smart, very well-paid product teams spend considerable time optimizing it.

Confirm-shaming: Making You Feel Bad About Saying No

Confirm-shaming is smaller than the Roach Motel but more personally offensive.

It works like this. A popup appears offering you something — a newsletter, a discount, an app download. The "yes" button says something enthusiastic: "Yes, I want to save 20%!" The "no" button, instead of saying "No thanks" or simply "Close," says something designed to make you feel faintly stupid or morally deficient for declining.

"No thanks, I prefer to pay full price." "No thanks, I don't care about my health." "No thanks, I already know everything about marketing." "No thanks, I hate saving money."

These are real examples, collected from real websites. The technique works — it measurably increases conversion rates — which is why it persists despite being, by any reasonable assessment, a minor act of psychological aggression toward the user.

The underlying mechanism is simple: it reframes declining as an admission of something undesirable, making acceptance feel like the rational choice. It's manipulative in the most literal sense — it's changing your behavior through psychological pressure rather than genuine persuasion.

What's particularly galling is the craft involved. Someone sat down and wrote those button labels. Someone else tested them. Someone approved them. At every step, a person with design skills made a deliberate choice to make users feel bad about exercising a preference. This is not a dark pattern that emerged from negligence. It is a dark pattern that required work.

Hidden Costs and the Checkout Sting

You've found the item. You've checked the price. It's reasonable. You add it to your cart, you proceed through three pages of checkout, you enter your card details, and then — on the final confirmation screen, the one that's one click away from completion — the price is different.

There's a "service fee." A "processing fee." A "facility fee." A "convenience fee" (for the convenience, presumably, of being allowed to give them money). Taxes that were not included in the displayed price, calculated at a rate that seems to bear no relationship to anything you recognize from adult life.

The total is now 40% higher than the price you thought you were paying.

This is Hidden Costs, and it is so pervasive in the ticketing, travel, and hospitality industries that most adults have simply accepted it as a feature of modern commerce. We have been conditioned to expect that the price shown is not the price paid, and we've mostly stopped being angry about it, which is itself a kind of dark pattern victory — the normalization of deception as just how things work.

Ticketmaster became so notorious for this that "Ticketmaster fees" entered the cultural lexicon as a shorthand for egregious hidden costs. Their typical fee structure, applied to a mid-range concert ticket, routinely adds 30–40% to the face value. A 2022 congressional hearing on the matter produced one of the more memorable exchanges in recent legislative history, when a senator asked a Ticketmaster executive to justify a $23 "facility charge" on a ticket to a venue that Ticketmaster did not own or operate.

The executive did not have a satisfying answer. Ticketmaster continued to charge facility fees.

The Pre-Ticked Box and the Consent Illusion

In the pantheon of dark patterns, the pre-ticked checkbox holds a special place, because it's the one that masquerades most convincingly as legitimate design.

Here's the scenario. You're completing a registration form. At the bottom, above the submit button, there are three checkboxes. Two of them are unchecked: you have to actively opt in to the newsletter and to sharing your data with "trusted partners." The third — let's say it's enrollment in auto-renewal — is pre-checked. You have to notice it, understand what it means, and actively uncheck it to opt out.

Most users do not notice it. This is not a hypothesis. It has been tested, repeatedly, and the data is consistent: pre-checked boxes dramatically outperform opt-in boxes in enrollment rates, not because users prefer the thing being offered but because the cognitive load of a long form causes them to miss it.

This is the consent illusion: a technical opt-in — you had the checkbox right there — that functions in practice as covert enrollment. The form satisfies a legal requirement while violating the spirit of it entirely.

GDPR, Europe's data protection regulation, was supposed to fix this. And in Europe, for the most part, it did make pre-ticked consent boxes illegal for data collection. The result was a generation of cookie consent banners that are themselves a masterclass in dark patterns — designed to make "Accept All" visually prominent and "Manage Preferences" buried, grey, and technically functional while being practically inaccessible.

Regulatory compliance and user-hostile design are not mutually exclusive. Some of the most sophisticated dark patterns in existence today were designed specifically to satisfy legal requirements while maintaining maximum extraction. This is a design discipline. A depressing one, but a discipline.

Enter AI: Dark Patterns with a Brain

Everything described so far is essentially static. Pre-designed friction, pre-written shame, pre-calculated obfuscation. The dark patterns of the early web were blunt instruments — they worked on everyone equally, crudely, at scale.

AI changes this in a way that should genuinely concern you.

The new generation of AI-powered dark patterns are not static. They are **adaptive**. They learn what works on you, specifically, and adjust in real time.

Consider dynamic pricing, which sounds neutral until you look at it closely. An airline's pricing algorithm knows — based on your browsing history, your device type, your location, the time of your search, and various inferred characteristics about your income and urgency — roughly how much you're willing to pay for a given flight. It can price accordingly. The same seat, at the same time, for a different price depending on who's asking.

This is not hypothetical. It is documented and widespread in travel, retail, insurance, and financial services. And the user has essentially no way to know it's happening, because the system presents a single price as if it's the price, rather than a price calculated specifically for them.

Or consider engagement optimization — the AI system, across almost every major social platform, that has learned with extraordinary precision what emotional triggers keep you scrolling. Not "content you enjoy" in any meaningful sense. Content that activates the specific psychological response — outrage, anxiety, social comparison, FOMO — that keeps your thumb moving and your attention on the screen.

This is not a recommendation engine. It is a behavioral manipulation engine, optimized against your own interests, running continuously, learning from every interaction you have with it.

The scale of this is hard to hold in your head. These systems are influencing the attention and emotional states of billions of people, simultaneously, in ways those people cannot see or opt out of. And they are getting better at it every year.

The Algorithmic Dark Loop

There is a failure mode worth naming here — one researchers call **mode collapse**— where users learn to work the system rather than doing genuine work. In the context of dark patterns, this has a consumer-facing equivalent I think of as the **dark loop**: a system designed not to serve you but to capture you, using your own behavioral data as the mechanism.

The dark loop works like this. You arrive. The system reads your state — your mood, your attention, your vulnerabilities. It serves you something calibrated to that state. You engage. The system logs the engagement and updates its model. It serves you something slightly more calibrated. You engage more. The loop tightens.

The exit from this loop is genuinely difficult to find, not because you lack willpower but because the loop has been engineered by teams of intelligent people with access to more data about your psychology than you have about yourself.

Instagram, in internal research that became public as part of the Facebook Papers in 2021, found that the platform's recommendation systems were pushing young users — teenage girls specifically — toward content about body image, dieting, and social comparison, not because users asked for it but because engagement data showed it triggered higher interaction rates. The system had learned that anxiety drives engagement. It optimized accordingly.

Nobody designed "make teenage girls feel bad about their bodies." But the optimization target, combined with behavioral data and no ethical guardrails, got there anyway.

The Designer's Complicity Problem

Here is the part of this chapter that is uncomfortable if you are, like me, a UX designer.

Dark patterns don't design themselves.

Every roach motel, every confirm-shaming button, every pre-ticked box, every notification optimized for emotional vulnerability was designed by a person. Often a talented person. Often a person who went to a good school, cares about design, uses the word "user-centered" in their portfolio, and goes home at night to a family they love.

The industry has spent decades building a professional identity around advocacy for users. UX designers are, in our own self-image, the people in the room who speak up for the human on the other end of the screen. We are user advocates. It says so on our LinkedIn profiles.

And yet. The dark patterns are there. And someone designed them.

The honest answer is that dark patterns usually don't arrive fully formed with a label on them. They arrive as "optimizations." As "conversion improvements." As "reducing friction in the retention flow." As A/B test results that show option B performs 12% better and nobody in the meeting asks "performs better for whom?"

The designer who labels a cancel button "Cancel Anyway" probably did not think of themselves as doing something harmful. They were solving a business problem. The problem was churn. The solution was friction.

The user was, in that framing, an obstacle rather than a person.

This is how it happens. Not in one dramatic ethical failure but in a series of small, locally reasonable decisions that add up to a system that is working against the people using it.

The professional responsibility here is real. Recognizing a dark pattern is the first step. Naming it in the room — "this is designed to frustrate users into not cancelling, and I'm not comfortable with that" — is harder. Refusing to ship it is harder still, because you are one person in a meeting and the business case is sitting right there in the A/B data.

But it is the job. It was always the job. And in the age of AI, where the dark patterns have gotten smarter, more adaptive, and more powerful, it is more the job than ever.

You are probably in a dark pattern if: you can sign up in one click but cancellation requires a phone call during business hours. If the "decline" option on a consent dialog is in light grey, very small, and nowhere near the "accept" button. If a price changed between the product page and the checkout confirmation. If a button says something other than "no" for what is, functionally, a no. If you feel vaguely guilty or foolish for declining something. If the app asked for a permission — location, contacts, microphone — that has no plausible connection to its stated purpose. If the notification settings have thirty-seven options and "turn off all notifications" is not one of them.

None of these are accidents.

Chapter 6: Trust, Transparency, and the Black Box

In which we discover that trust is not a feeling but a set of beliefs, that explaining yourself and being understood are completely different things, and that "trust us, it's complicated" is not a design strategy.

In 2017, a federal court ruled that the Houston Independent School District had violated teachers' due process rights by using a proprietary algorithm called EVAAS to evaluate — and in some cases fire — teachers whose inner workings could not be examined or challenged.

The teacher's union tried to understand why — what inputs had produced that score, what weighting had been applied, what data the system had used — the district told them the algorithm was proprietary. The vendor declined to explain it. There was no mechanism to appeal. There was no way to see inside the box.

A federal court later ruled that the Houston Independent School District had violated teachers' due process rights by using a system whose inner workings could not be examined or challenged. But the ruling came years after the firings.

This is the black box problem. And it is not going away.

Side note: *When researching this book, I wanted to cite this example, but I wanted more details about the Houston School District. I asked a Chatbot to help me do some research. This is what it wrote:*

'In 2016, a teacher in Houston named Sabrina Demming received an evaluation from her school district's new performance management system. The system was called EVAAS — Education Value-Added Assessment System — and it used a proprietary algorithm to measure each teacher's contribution to student achievement. Demming had been rated highly by her principal. Her students liked her. Her colleagues respected her. The algorithm gave her a score of zero. She was fired.'

This was interesting, so I wanted to research poor Sabrina and put her in the book. There was one problem. Sabrina doesn't exist. Or at least I couldn't find her. I asked the Chatbot for clarification. It replied:

'I cannot find a verified reference to Sabrina Demming specifically. The name and the personal story as I wrote it — her principal's rating, her score of zero, her firing — I cannot confirm is accurately attributed. It's possible I generated a plausible-sounding specific name and narrative around a real general situation, which is exactly the kind of hallucination you are writing about warns about. That would be embarrassing in print.'

We all make mistakes, but writing about AI, and using AI, I am acutely aware of the traps. The authoritative way it wrote about poor Sabrina was very persuasive.

And a warning.

What Trust Actually Is

Trust, in this context, is a set of beliefs across three dimensions: **competence** (can this system do what it claims?), **predictability** (will it behave consistently?), and **alignment** (is it working toward my interests, or someone else's?).

You can trust a system on one dimension and not another. You might believe a recommendation engine is competent — it has good taste, technically — while doubting its alignment, because you suspect it's recommending things you'll buy rather than things you'll love. You might believe a medical AI is well-intentioned while doubting its predictability, because it produces inconsistent outputs that are hard to verify.

All three dimensions matter. And AI systems, as currently designed, have systematic weaknesses in all three.

Competence is uncertain because, as we discussed in Chapter 3, fluency is not accuracy. Predictability is undermined by the probabilistic nature of these systems — the same input can produce meaningfully different outputs on different days, which is deeply alien to users accustomed to deterministic software. And alignment is, to put it gently, not always what it appears.

The Explainability Trap

The first instinct, when users don't trust an AI system, is usually to explain it.

Add a tooltip. Write a help article. Include a "why did I see this?" link. Show the model's reasoning. Make the black box transparent.

This instinct is understandable. It is also, in most cases, wrong — or at least incomplete.

Here's the problem. Real explainability in modern machine learning systems is extraordinarily difficult. A deep neural network making a recommendation or a prediction is not following a logical chain of reasoning that can be rendered as a sentence. It's the product of billions of weighted parameters, trained on vast datasets, producing outputs that emerge from a process that even their creators cannot fully trace. Asking an AI system to explain itself in human terms is a bit like asking a river to explain why it chose this particular path to the sea.

What most "explanations" in AI products actually are is a post-hoc rationalization — a human-readable story constructed after the fact to make the output seem reasonable. The story may be plausible. It may even be partially accurate. But it is not a genuine account of why the system produced that output. It's a translation, at best. At worst it's a fiction that gives users false confidence in understanding something they don't.

This matters because users who believe they understand a system will stop questioning it. And a system that is questioned less is a system whose errors go uncaught longer.

Evidence, Not Explanation

I love the quote that cuts right through this: *"Evidence beats explanation."*

It's worth sitting with that for a moment, because it reframes the whole problem.

Users don't need to understand how a system reached its conclusion. They need to know whether to act on it. Those are different requirements, and they call for different design responses.

Explanation tries to open the black box. Evidence sidesteps it entirely — it gives users what they need to make their own judgment, without requiring them to trust the box at all.

What does evidence look like in practice?

It looks like the medical AI that shows you which regions of the scan it flagged, so the radiologist can look at those regions and form their own view. It looks like the legal research tool that shows you the source documents alongside its summary, with the relevant passages highlighted. It looks like the financial forecasting system that shows you the historical accuracy of similar predictions, so you can calibrate how much weight to give this one. It looks like the recruitment tool that shows you the specific experience and qualifications it matched, rather than just a suitability score.

In each case, the system is not explaining itself. It is showing its work, in a form the user can actually evaluate. The user remains the judge. The system is providing inputs to that judgment, not replacing it.

This is a fundamentally different design philosophy from the black box with a help article attached, and it produces fundamentally different outcomes in user trust and error-catching rates.

The Confidence Score Problem

At some point in the last decade, someone decided that what AI systems needed to communicate uncertainty was a confidence score. A percentage. "87% match." "Confidence: high." "Relevance score: 0.94."

This was well-intentioned and is mostly counterproductive.

The problem is not with quantifying uncertainty in principle. The problem is that users interpret numerical confidence scores as objective measurements of accuracy, when they are actually internal model metrics that don't map cleanly onto real-world correctness. A model that says "87% confident" is not telling you it will be right 87% of the time. It's telling you something about the distribution of its training data and the characteristics of this particular input — which is a different thing, and a thing that requires statistical literacy to interpret correctly.

Studies on how users respond to confidence scores consistently show two failure modes. Users with low statistical literacy tend to treat any high-confidence score as a green light, regardless of the actual error rate. Users with high statistical literacy sometimes become more confused by scores that don't match their domain expertise. Neither group is well served.

The alternative — showing uncertainty through ranges, alternatives, and comparative examples rather than scores — is harder to design but more genuinely useful. "This prediction falls within the range of similar past cases that turned out to be X" tells you something actionable. "Confidence: 87%" tells you something that feels actionable but probably isn't.

The Right to Understand

The Houston teachers case is an extreme example, but the principle it exposes applies across an enormous range of AI deployments: systems that make or influence significant decisions about people's lives, with no mechanism for those people to understand or challenge the decision.

Credit scoring. Insurance pricing. Bail recommendations. Job application screening. Content moderation. Medical triage. Benefits eligibility. These are not niche applications. They are decisions that shape people's opportunities, freedoms, and livelihoods — and in a growing number of cases they are being made, or heavily influenced, by systems whose reasoning is opaque to everyone except, sometimes, their vendors.

The legal concept that is slowly emerging to address this is sometimes called "algorithmic accountability" — the idea that when an automated system makes a consequential decision about a person, that person is entitled to a meaningful explanation of how that decision was made, and a meaningful path to challenge it if it's wrong.

The EU's AI Act, passed in 2024, goes further than most, requiring that high-risk AI systems maintain documentation, enable human oversight, and provide transparency to affected individuals. It is imperfect legislation about a fast-moving target, and it will be argued about by lawyers for years. But the underlying principle it's reaching for is correct: **consequential decisions require accountable processes.**

This is not just a legal question. It's a design question. The interface is where accountability becomes real or doesn't. A legal requirement for explainability that manifests as a 4,000-word technical document available on request is not meaningful transparency. Meaningful transparency is an interface that shows a person, in terms they can understand and act on, why a decision was made and what they can do about it.

Nobody does this perfectly yet. Some are doing it better than others.
It is one of the most important unsolved problems in UX right now.

Accessibility

I once worked for the Dutch Department of Health. They had an entire department dedicated to supporting the interests of blind citizens — and every person working in it was blind. There were no screens. No monitors, no visual interfaces of any kind. Just braille bars and screen readers. They experienced the web entirely through sound — websites read aloud, navigation announced, content linearized into audio by software that interpreted whatever the designer had built, or failed to build, for that purpose.

What struck me then, and has stayed with me since, is how unforgiving that experience was of design decisions made without those users in mind. An unlabeled button. An image with no alt text. A modal dialog that trapped the focus of a screen reader and couldn't be escaped. Each of these is an inconvenience for a sighted user. For a blind user it is a wall.

AI is making this both better and significantly more complicated. On the better side: AI-powered tools can now generate alt text automatically, describe images in real time, and make previously inaccessible content navigable. These are genuine advances and they matter.

On the complicated side: AI-generated interfaces introduce a new category of accessibility failure. A system that generates UI dynamically — that creates layouts, populates content, and adjusts presentation in real time based on behavioral data — can produce outputs that no accessibility audit has ever reviewed, because the output didn't exist until the user requested it. The WCAG guidelines, the ADA compliance frameworks, the screen reader compatibility testing — all of these assume a relatively stable interface that can be examined and certified. A generative interface is moving too fast for that model.

The blind users in that Dutch department were, in a very real sense, the most honest test of whether a website worked. They had no visual cues to compensate for structural failures. The design either communicated clearly, or it didn't. AI has inherited that test and is currently failing significant parts of it — not through malice but through the same default that produces every other accessibility failure: designing for the assumed majority and treating everyone else as an edge case.

They are not an edge case. They are users. The same argument this book has been making throughout applies here with particular force: there is a person on the other end of this screen. Sometimes that person is experiencing your design in a completely different way than you imagined. Design like it.

Transparency Theatre

There is a version of transparency that looks like transparency and isn't.

You've seen it. Every major platform, since the wave of regulatory pressure in the late 2010s, now has a privacy dashboard. Settings pages. Preference centers. Data download options. "Why did I see this ad?" buttons that produce answers of such magnificent vagueness as to be essentially decorative.

Facebook's "Why am I seeing this ad?" feature, for example, has at various points explained ad targeting with reasons like "you're in the target audience this advertiser chose" — which tells you literally nothing — or cited interest categories so broad ("technology," "travel") that no useful inference could be drawn from them about what data was actually used.

This is transparency theater: the performance of openness without the substance. It exists because it satisfies the letter of regulatory requirements, appeases users who feel vaguely surveilled but haven't thought too hard about it, and provides the platform with a defense against accusations of opacity. "We show users exactly why they see each ad." Technically true. Practically meaningless.

The tell for transparency theater is whether the transparency actually changes anything. If showing users their data leads to no mechanism for them to meaningfully control it, the transparency is decorative. If the explanation of a decision provides no route to appealing it, the explanation is cosmetic. Real transparency is transparency that does something. That gives you a lever. That makes a difference.

This has been described to me as "transparency with teeth" — provenance, user control, and audit trails that are real rather than ceremonial. That phrase should be a design principle. Not "does this explain the system?" but "does this explanation actually give the user anything?"

When the Machine Changes Its Mind

There is a specific trust problem in AI systems that has no real equivalent in traditional software, and it is one that most designers haven't fully grappled with yet.

Models change.

Not because anyone pressed a button and made them different, but because the underlying model was updated, or the training data was refreshed, or a parameter was adjusted in response to new findings. These are often improvements — the system is more accurate, more fair, more robust. But from the user's perspective, the experience is unsettling: a system that behaved one way yesterday behaves differently today, for no visible reason.

Consider a medical professional who has spent six months learning the patterns of an AI diagnostic tool. They've built an intuition for when to trust it and when to push back. They've calibrated their own judgment against its outputs. Then the model is updated — quietly, without announcement — and the patterns shift. Their calibration is now wrong. The trust they've built was trust in a system that no longer exists.

This is not hypothetical. It happens in enterprise deployments regularly. And the design response — "we updated the model to improve accuracy" in a changelog buried in a help center — is not adequate.

Users who form working relationships with AI systems deserve to know when those systems have materially changed. Not at a technical level — nobody needs to read the code — but at the level of behavior: *this system may produce different outputs than it did previously in the following categories.* That's a sentence that respects the user's investment in understanding the tool.

The alternative is a relationship built on sand — where users think they understand something that keeps silently changing under them, and every unexplained behavioral shift erodes a little more of the trust they've extended.

Designing for the Sceptic

One of the more useful reframes in designing trustworthy AI experiences is to design not for the credulous user but for the skeptic.

The credulous user will accept whatever the system tells them. They are well-served by a confident, clean interface. They are also the user who will follow AI-generated medical advice without checking it, sign a contract because the AI said it looked fine, and hire a candidate because the algorithm scored them highest. These are not good outcomes, even when the AI happens to be right.

The skeptic wants to check. They want sources. They want to understand the basis for a claim. They want to know what the system doesn't know. They are, in terms of outcomes, the user you want more of — because their skepticism is the error-catching mechanism that compensates for the system's inevitable imperfections.

Good transparent design rewards skepticism. It makes verification cheap, not laborious. It gives the skeptic what they need to check without forcing them to fight the interface to do it. It treats "I'd like to look at the underlying data" not as an edge case but as a legitimate and valuable user behavior.

The irony is that designing for the skeptic also serves the credulous user better. A system that shows its sources, shows its uncertainty, and makes correction easy is a safer system for everyone — including the people who never use any of those features.

The Trust Recovery Problem

We established in Chapter 3 that trust withdrawals are larger than trust deposits, and that algorithm aversion — the tendency to abandon AI tools after seeing a single visible failure — is a documented and durable phenomenon.

What we haven't discussed is what it takes to get trust back.

The research here is not especially encouraging. Trust recovery after a visible AI failure is slow, asymmetric, and heavily dependent on what happens immediately after the failure. Systems that acknowledge errors clearly, explain what went wrong in user terms, demonstrate that the error has been addressed, and give users meaningful control going forward recover faster than systems that do none of those things. Which sounds obvious until you look at how most enterprise AI deployments actually handle visible failures: with silence, or with a support ticket process, or with a product update note six weeks later that mentions "improved accuracy" without acknowledging what was inaccurate.

The fastest path to trust recovery is also the simplest, and the most consistently avoided: **honesty in real time.** Not "we take accuracy very seriously." Not a help article. The interface, at the moment of the error, saying something like: *we got this wrong. Here's what we think happened. Here's what you can do now.*

That requires institutional courage as well as design skill. It requires someone deciding that user trust is worth more than the optics of admitting failure. In my experience, that decision is harder to get made in a product meeting than almost any design decision I've encountered. But the products that make it consistently are the ones that build durable relationships with their users rather than just good first impressions.

A Word on the Black Box We Keep

There's one more transparency problem worth naming, and it's the one closest to home.

Designers know things about their users that those users don't know are being collected, stored, analyzed, or acted on. We build experiences around behavioral data that users have nominally "consented" to in terms of service agreements that no human has ever voluntarily read cover to cover. We A/B test on real users without telling them. We use session recordings and heatmaps that capture every mouse movement and hesitation.

Most of this is legal. Much of it is genuinely useful for making better products. Some of it is surveillance that we've gotten comfortable calling research.

The transparency we demand of AI systems — show your sources, explain your reasoning, make your assumptions visible — is transparency we should be willing to apply to our own practice. The same question we should ask of every AI feature — *what would this user think if they knew exactly what this system was doing with their data and behavior?* — is a question designers should ask of themselves.

It's an uncomfortable question. It's supposed to be.

Chapter 7: Anxiety, Dependency, and the Atrophied User

In which we discover that the brain is use-it-or-lose-it, that convenience has a hidden invoice, and that the most important question about any technology that does something for you is what happens to you when it doesn't.

In 2011, a team of neuroscientists at University College London published a study that got more attention in the popular press than most neuroscience research manages. They had been studying London taxi drivers — specifically, the part of the brain called the hippocampus, which plays a central role in spatial navigation and memory formation.

What they found, building on earlier research, was that experienced London cab drivers had measurably larger hippocampal volumes than control subjects. Years of navigating one of the world's most complex street networks — learning "The Knowledge," the extraordinary memorization feat required to earn a London taxi license — had physically enlarged the relevant brain structure. The brain had grown to meet the demand placed on it.

Then the researchers looked at what happened after GPS became widespread in the taxi industry.

The growth stopped. In some cases, it reversed.

The technology hadn't just changed how drivers navigated. It had changed their brains.

The Use-It-Or-Lose-It Brain

Neuroplasticity — the brain's ability to reorganize itself in response to experience — is one of the more remarkable things about human cognition. The brain is not fixed hardware running fixed software. It is a dynamic system that continuously remodels itself based on what you ask it to do.

Ask it to navigate complex environments repeatedly, and the navigation structures strengthen. Ask it to remember phone numbers, and it builds capacity for that. Ask it to read music, recognize faces, throw a ball, speak a second language — all these activities leave measurable traces in brain structure and function.

The corollary is equally true, and considerably less comfortable: **stop asking the brain to do something, and the relevant capacity diminishes.**

This is not a moral failing. It's basic neuroscience. The brain is metabolically expensive to run, and it does not maintain infrastructure it isn't using. Circuits that go quiet get pruned. Skills that go unpracticed degrade. The brain that offloads navigation to a GPS app is a brain that is, over time, investing that freed-up capacity elsewhere — or simply not investing it at all.

For most of human history, this trade-off was manageable because the things technology replaced were genuinely tedious and low-value. Calculators replaced mental arithmetic. Spell-checkers replaced the laborious process of looking up words. Spreadsheets replaced accountants doing sums by hand on paper. In each case, something was offloaded, but the offloaded thing was not something we particularly needed to be good at as humans.

The question now is whether we're still in that category of trade-offs, or whether we've crossed a line into offloading things that actually matter for human capability, judgment, and wellbeing.

I think we've crossed the line. And I think we did it quietly, without noticing, over the course of about a decade.

The Spell-Checker Was Just the Beginning

Let's take writing, since it's a skill most knowledge workers rely on professionally.

Spell-checkers arrived in the 1980s and were, without question, a net positive. The ability to catch typos and common misspellings removed a source of friction from writing without touching the underlying skill. You still had to form sentences, develop arguments, choose words. The checker just caught the mechanical errors.

Grammar checkers were a slightly more complex case. Tools like Grammarly do more than catch errors — they suggest restructured sentences, flag passive voice, recommend word substitutions. Users who rely heavily on these tools report, anecdotally, a growing uncertainty about their own grammatical instincts. The tool has partially replaced the internal editor, and the internal editor — like the London cabbie's hippocampus — has started to shrink from disuse.

Now consider what AI writing tools do. They don't just check your writing. They generate it. You provide a prompt — a rough idea, a brief, sometimes just a topic — and the system produces prose. You review it, perhaps edit it, perhaps just send it.

This is genuinely useful. I am not going to pretend otherwise, having used these tools myself and found them valuable for specific tasks. But there is a question lurking behind the usefulness that the industry has no interest in asking, because the answer is bad for business: **what is happening to the writing ability of people who use these tools heavily over years?**

We don't fully know yet, because the tools are too new and the longitudinal research hasn't caught up. But the neuroplasticity evidence suggests a hypothesis. If writing — the struggle to find the right word, to construct a coherent argument, to organize thought into language — is a cognitive workout, and you systematically replace that workout with an AI-generated output, the relevant cognitive muscles will, over time, weaken.

The person who outsourced their writing to an AI for three years and then has to write something important — without the tool, under pressure — will find themselves in the position of the pilot who hasn't hand-flown in eighteen months and is suddenly asked to land the plane in a crosswind.

The Surgeon Who Stopped Operating

There's a concept in cognitive science called **skill atrophy**, and it is about to become one of the defining anxieties of the next twenty years.

Here's the uncomfortable version of the GPS story I told at the beginning.

Studies have repeatedly shown that heavy reliance on navigation apps degrades spatial memory. People who use GPS constantly are measurably worse at forming and retaining mental maps of areas they've visited frequently. Their brains are not being lazy — they've simply offloaded a task to an external system and, as brains do, have efficiently stopped maintaining the now-unnecessary circuitry.

This is not a problem when you're navigating a city. It becomes more interesting when the skill being offloaded is, say, making a clinical diagnosis. Or landing an aircraft. Or reading a legal contract.

There's a famous aviation case study that gets cited regularly in human factors research: as cockpit automation increased, pilots became measurably worse at hand-flying aircraft in emergency situations — precisely because they rarely needed to do it. The automation that made aviation safer in normal conditions created a class of risk in abnormal ones.

The design question this raises is not "should we automate this?" The design question is: **what happens to the human when we do?**

The best AI UX designs I've seen take this seriously. They build in what researchers call *cognitive forcing functions* — small, deliberate moments of friction that keep users engaged and thinking, rather than just approving. A medical AI that shows you its reasoning and asks you to confirm before acting. A legal tool that highlights the three passages you most need to read before the summary. A navigation app that occasionally, briefly, asks you to decide — not because it can't, but because you should.

This is not making life harder. This is keeping you good at your own job.

The Pilot Problem, Revisited

The aviation case keeps coming up in discussions of AI dependency, and for good reason. It is the most extensively studied example of skill atrophy in a high-stakes professional context, and the findings are not reassuring.

As commercial aviation automated more and more of the flying — autopilot, autothrottle, automated approach and landing systems — accident investigators began noticing a pattern in the incidents that did occur. A disproportionate number involved situations where automation had unexpectedly disengaged, and pilots had failed to respond appropriately. Not because they were incompetent. Because they were out of practice.

The aviation industry's response has been instructive. Rather than pulling back on automation — which genuinely does make aviation safer in normal conditions — they mandated periodic manual flying requirements. Pilots are now required to regularly fly without automation, specifically to maintain the skills that the automation has otherwise made unnecessary. Deliberate practice of an atrophying skill, mandated by regulation.

This is the design insight that the consumer technology industry has almost entirely failed to apply. The question is not "should we automate this?" The question is "if we automate this, how do we ensure the human doesn't lose the underlying capability?"

Most products have no answer to that question, because maintaining user capability is not a metric that appears on any dashboard, and the short-term experience of automation is always better than the short-term experience of doing it yourself. The trade-off is invisible until it isn't.

Dependency by Design

There's a version of this problem that is accidental — a side effect of useful tools, unintended, worth addressing but not malicious. And then there's a version that is deliberate.

Platform dependency is one of the oldest and most profitable strategies in technology. The goal is not simply that users use your product. The goal is that users *need* your product — that switching away becomes practically, cognitively, or socially expensive. Lock-in is the polite industry term. Dependency is the accurate one.

AI tools have introduced a new and particularly effective form of dependency: **workflow integration dependency**. As AI assistants become embedded in how people do their jobs — drafting communications, organizing information, summarizing meetings, generating first drafts of everything — they become load-bearing elements of professional practice. Removing them doesn't just make you slower. It makes you structurally unable to operate at the same capacity, because the capacity was being provided by the tool.

This is different from, say, being dependent on email. Email is infrastructure. It doesn't think. Dependency on an AI that thinks — that makes judgments, drafts outputs, filters information — is dependency on a cognitive prosthetic. And cognitive prosthetics, unlike email, can be withdrawn, changed, or monetized in ways that affect the quality of the thinking itself.

The enterprise software industry has understood this for decades. Systems like SAP are difficult and expensive to replace not primarily because they are technically superior but because organizations have built their entire operational logic around them. The switching cost is the product.

AI tools are building the same moat, faster, at the level of individual cognitive practice rather than organizational infrastructure. The person whose entire professional output flows through a single AI platform is a person whose career is, in a meaningful sense, dependent on that platform's continued existence, goodwill, and pricing decisions.

This should feel uncomfortable. It is supposed to.

The Anxiety Loop

So far we've talked about dependency in terms of capability — what you can no longer do without the tool. But there's a parallel problem in terms of emotional state — what you feel without it.

Notification anxiety is well documented and widely experienced. The feeling of unease when you haven't checked your phone in a while. The compulsive glance at the lock screen. The low-level discomfort of being unreachable, or of potentially having missed something. This is not a personality quirk. It is a learned response, carefully cultivated by notification systems designed to create exactly this feeling — because users who feel anxious about missing things check more often, and users who check more often generate more engagement data and more ad impressions.

The anxiety is the product.

AI tools are developing their own anxiety loop, and it's worth watching. As these tools become more capable, users who use them heavily begin to experience a particular kind of performance anxiety without them. Not just "this will take longer without the tool" but something more fundamental: *am I actually good enough to do this without it?*

This is a new psychological state, and it's one that the technology industry has, characteristically, not thought much about. Users who begin to doubt their own judgment, their own writing, their own analytical capability — because those capabilities have been consistently augmented and they're no longer sure where the augmentation ends and they begin — are users in a fragile relationship with technology.

They're also users who are very unlikely to stop using the tool. Which is, from a retention perspective, an excellent outcome. From a human perspective, it is something closer to a trap.

What the Research Actually Shows

The dependency literature is more nuanced than the alarm-bell version suggests, and it's worth being accurate about it.

Not all offloading is harmful. Cognitive offloading — using external tools to extend your mental capacity — is a normal and healthy part of human cognition. Writing itself is cognitive offloading. So are calendars, to-do lists, and the notes you take in meetings. The idea that you must hold everything in your head to truly know it is not a finding of cognitive science. It's a romanticization of difficulty.

The research distinguishes between offloading that **extends** capability and offloading that **replaces** it. Using a calculator to handle arithmetic you couldn't do in your head extends your capability. Using a calculator so consistently that you lose the ability to do arithmetic you once could do in your head is replacement. The distinction matters.

What the research consistently flags as problematic is offloading in domains where the struggle itself is the learning — where the difficulty of the task is not a bug but the mechanism by which competence is built. Writing is one of these domains. Clinical reasoning is another. Legal judgment, strategic thinking, creative problem-solving — anywhere that the process of working something out is what builds the underlying skill, replacing that process with an AI output removes the training stimulus.

The design implication, again, is not "don't use AI." It is: **be honest about what the AI is doing to the human who uses it over time, and design accordingly.** Build in the manual flying practice. Don't just optimize for the smoothest assisted experience.

The Productivity Paradox

Here's one that tends to produce arguments at dinner parties.

AI tools are almost universally described as productivity enhancers. They save time. They reduce effort. They enable individuals to produce more output with less input. Every product in this space leads with this claim, and in the short run it is largely true.

But productivity at what, exactly?

If an AI writing tool enables you to produce three times as many words per hour, and those words are adequate but not distinctively good, you have increased the quantity of your output while potentially capping the ceiling of its quality. The tool has made you faster at average work. It has not made you better at exceptional work — and it may, over time, be degrading the capability that exceptional work requires.

This is the productivity paradox of AI tools: they are most useful for work that doesn't benefit from the full depth of human judgment, and least useful for work that does. Which is fine, until the boundary between those categories starts to shift — and users, habituated to the tool, start applying it in domains where it shouldn't be trusted.

The researcher who uses AI to summarize papers they should have read. The doctor who uses AI to draft clinical notes and gradually stops doing the cognitive work of forming independent observations. The lawyer who uses AI to research precedents and loses fluency in the primary sources. In each case, the productivity gain is real and immediate. The capability loss is slow, invisible, and potentially severe.

Designing for Capability, Not Just Convenience

I want to be clear that this chapter is not a Luddite argument. I am not suggesting we return to paper maps, longhand writing, and manual arithmetic in the name of keeping our hippocampi large.

What I am suggesting is that designers — and the organizations that employ them — have a responsibility that the industry has not yet taken seriously: **designing for the long-term capability of users, not just the short-term smoothness of their experience.**

This looks like a few specific things in practice.

It looks like **progressive disclosure of automation** — starting users with more manual control and introducing automation incrementally, so they build the underlying understanding before the shortcut. The cooking analogy: learn to make the sauce before you use the jar. The jar is fine. But you should know what's in it.

It looks like **visible skill maintenance** — building moments into AI-augmented workflows where users are asked to do something themselves, not because the AI can't but because the practice matters. The mandatory manual flying. The occasional "what do you think before I show you what I found?" moment.

It looks like **honest capability mapping** — telling users clearly what a tool does and does not replace, so they can make informed decisions about where they want to offload and where they want to retain the skill. Not "AI does the hard work for you." But "AI handles X so you can focus on Y — and here's what that means for your development in X over time."

None of this is impossible. Some of it is already happening, in thoughtful products built by designers who are asking the right questions. But it is not the default. The default is to optimize for the immediate experience, ship the feature, and let the long-term human consequences be someone else's problem.

They are not someone else's problem. They are the problem. They are what this book is about.

The Question Worth Asking

There is one question I have started asking about every AI feature I encounter, in my own work and in the products I use:

If this tool disappeared tomorrow, would I be better or worse at this than I was before I started using it?

For some tools, the answer is clearly better — the tool has been teaching me, extending me, making me more capable in ways that would survive its removal. Those are good tools. Those are tools designed with the human in the loop in the right way.

For others, the honest answer is worse. I have offloaded something I used to do, the offloading has been convenient, and I have not maintained the underlying skill. Those tools have made me faster and more fragile simultaneously.

The ratio of the first kind to the second kind is, in my experience, not encouraging.

But it is a ratio that design can shift. That is the point. That is what the rest of this book is building toward.

Chapter 8: Control, Consent, and the Illusion of Choice

In which we discover that a settings page with forty-seven options is not the same as control, that consent given under duress to a document nobody read is not really consent, and that the most effective cage is the one that looks like a room.

In 1971, a psychologist named Ellen Langer ran a series of experiments that produced one of the more unsettling findings in the history of behavioral research.

She gave people a lottery ticket. Half of them were handed a ticket at random. The other half were allowed to choose their own ticket from a selection. Then, before the lottery was drawn, she offered to buy the tickets back.

The people who had chosen their own tickets demanded significantly more money to give them up than the people who had been handed theirs — even though both tickets had identical odds of winning, and the act of choosing had conferred no actual advantage.

Langer called this the **illusion of control**: the tendency for people to behave as though they have influence over outcomes that are, in fact, determined by chance or by forces entirely outside their influence.

The technology industry read this research and, whether deliberately or through the emergent logic of product optimization, built an entire design philosophy around it.

Give people a choice. It doesn't matter much what the choice actually does. The **feeling** of choosing is what counts.

The Settings Page That Does Nothing

Open the settings menu of any major social platform and you will find something that looks, at first glance, like an impressive suite of user controls. Privacy settings. Data preferences. Ad personalization options. Notification granularity. Content filters. A full page — sometimes several full pages — of toggles and dropdowns that collectively imply: *you are in charge of this experience.*

Now turn everything off.

On most platforms, the practical difference between "full personalization" and "all personalization disabled" is marginal at best and invisible at worst. The algorithm still runs. The behavioral data is still collected. The ads are still shown — they're just, technically, "not personalized," which in practice means they're targeted by context and demographics rather than individual behavioral profile. The feed is still curated. The engagement optimization still operates.

This is not because the engineers forgot to connect the settings to the system. It's because the settings are, in many cases, not designed to give you meaningful control. They're designed to give you the *feeling* of meaningful control — and then to quietly route around your preferences at the system level, where you can't see it happening.

The FTC's 2022 report on commercial surveillance found, among many things, that a significant number of user controls in major platforms were either non-functional, misleadingly labeled, or designed in ways that made opting out of data collection technically possible but practically inaccessible. Dark patterns, in other words, applied specifically to the controls that were supposed to protect you from dark patterns.

If you find that recursive, you should. It's entirely intentional.

The Terms of Service Nobody Reads

In 2008, two researchers at Carnegie Mellon calculated how long it would take the average American to actually read the privacy policies of the websites they visit in a year. Their estimate: **76 work days.**

Not 76 days of casual reading. 76 full eight-hour working days, devoted entirely to reading privacy policies, just to be technically informed about the data practices of the services you use.

The situation has not improved since 2008. If anything, privacy policies have gotten longer. The average length of a major platform's terms of service document now runs to tens of thousands of words — longer than many novels — written in legal language calibrated to satisfy regulatory requirements while being comprehensible to approximately nobody who isn't a specialist in technology law.

This is consent theater at industrial scale.

The legal framework of informed consent — the idea that you have agreed to something by clicking "I Agree" — was designed for a world where agreements were readable, comprehensible, and roughly proportional in length to what was being agreed to. It was not designed for a world where agreeing to use a free photo-sharing app requires implicitly consenting to a 50,000-word document that grants the platform a perpetual, worldwide, royalty-free license to use your face in advertising in jurisdictions you've never visited.

No one reads these documents. The companies writing them know no one reads them. The lawyers drafting them know no one reads them. The regulators who theoretically enforce them know no one reads them. The entire apparatus of online consent is a legal fiction that everyone involved has agreed, by mutual unspoken compact, to treat as real.

The reason it persists is straightforward: the alternative — consent that is genuinely informed, genuinely voluntary, and genuinely specific — would be commercially catastrophic for the business models that depend on extracting maximum data with minimum friction. Meaningful consent would mean less data. Less data means less targeting. Less targeting means less ad revenue. The math doesn't work for the platforms, so the consent theater continues.

The Architecture of Manufactured Consent

Consent in digital products is not just poorly designed. It is, in many cases, actively designed against.

The cookie consent banner is the purest current example, and worth examining in some detail because it has become so normalized that most users have stopped registering it as a choice at all.

The cookie banner exists because European privacy law — specifically the ePrivacy Directive and subsequently GDPR — requires that users consent to non-essential cookies before they're set. This was a genuine regulatory win for privacy advocates. The intent was clear: users should be able to choose whether to be tracked.

The implementation was a masterclass in compliant non-compliance.

A well-designed cookie banner, in accordance with regulatory guidance, should present "accept" and "reject" options with equal visual prominence. The user should be able to decline all non-essential cookies in one click, just as they can accept them in one click.

What most cookie banners actually do: present "Accept All" as a large, prominent, colored button. Present "Manage Preferences" or "Reject Non-Essential" as small, grey, text-only links, sometimes requiring scrolling to find, sometimes hidden behind an additional modal, sometimes labeled in ways that make it unclear whether clicking them actually declines anything or simply opens another layer of options.

Studies of cookie banner design consistently find accept rates far above what genuine neutral presentation would produce. One 2022 study found that dark-pattern cookie banners achieved consent rates of up to 90%, while neutral designs — equal prominence, one-click rejection — produced consent rates closer to 30 to 40%.

The 50-percentage-point difference between those numbers represents the commercial value of the dark pattern. It is, at scale, worth billions of dollars in advertising revenue annually. And it is extracted specifically by exploiting the gap between the letter of consent law and the spirit of it.

Nudge Architecture and the Choice Environment

Not all of this is sinister. Some of it is a legitimate application of behavioral economics, and the distinction matters.

In 2008, Richard Thaler and Cass Sunstein published a book called *Nudge*, which argued that the architecture of choice environments — the way options are presented, ordered, and framed — inevitably influences decisions, and that architects of those environments have a responsibility to design them in ways that serve people's genuine interests. They called this **libertarian paternalism**: preserving freedom of choice while nudging people toward better outcomes.

The canonical example is organ donation default settings. Countries where organ donation is opt-out — where you're automatically a donor unless you explicitly remove yourself — have dramatically higher donation rates than countries where it's opt-in. The choice is free in both systems. The default is what changes. And the default does most of the work.

Thaler and Sunstein were talking about public policy. The technology industry absorbed the insight and applied it to product design — but with a crucial difference in whose interests the nudge serves.

A nudge toward organ donation serves the interests of patients who need organs, and arguably the interests of most donors who would, if asked reflectively, choose to donate. The interests of the nudge designer and the interests of the nudged are aligned.

A nudge toward accepting full data collection serves the interests of the platform. The interests of the nudge designer and the interests of the nudged are in direct conflict. This is not libertarian paternalism. It's just manipulation, dressed in the language of behavioral economics.

The difference is not subtle. But the design looks similar, which is why the "we're just optimizing the choice environment" defense persists in product discussions where it has no business being taken seriously.

The Paradox of Choice, Exploited

Barry Schwartz, in his 2004 book *The Paradox of Choice*, documented something that designers have known intuitively for decades: beyond a certain point, more options produce worse decisions and lower satisfaction. Too much choice creates cognitive overload, decision paralysis, and post-decision regret — the nagging sense that you might have chosen better if you'd chosen differently.

Good design has always used this knowledge to simplify and clarify — to reduce the option space to what's genuinely useful and present it in a way that helps people decide well.

Bad design uses it to exhaust people into choosing the default.

The forty-seven-toggle privacy settings page is not designed for users who want fine-grained control. It's designed to overwhelm users who want to opt out of something, producing exactly the cognitive overload that Schwartz documented, until they give up and click "Accept All" or close the tab without changing anything.

This is the exploitation of a genuine cognitive limitation — our limited capacity for complex, multi-variable decision-making — in the service of the platform's data extraction goals. The complexity is not a design failure. It is the design.

AI has made this worse in a specific way. As AI systems make more decisions about your experience — what you see, what you're offered, how your data is used — the number of variables that a genuinely informed user would need to understand and control has grown faster than any settings page could possibly accommodate. The gap between "technically you have control" and "practically you have control" has never been wider. And it is widening every year.

When AI Makes Choices Before You Arrive

Everything so far has been about the choices users are presented with. But there's a deeper layer of the control problem that is harder to see and harder to address: the choices that have already been made before the user ever appears.

Every AI-powered interface has made thousands of decisions before you arrive. What to show you. What to filter out. What order to present options. What language to use. What emotional register to set. What your preferences probably are, based on behavioral profiles of people like you. What your goals probably are, based on the same.

By the time you're looking at the screen, the choice environment has been almost entirely constructed. What feels like an open field of options is a carefully curated set, pre-selected by an algorithm optimized for something — and that something may or may not be what you'd choose to optimize for if you were making the choice explicitly.

This is not inherently sinister. Curation is necessary. An interface that presented every possible option, with no pre-selection, would be unusable. The question is always: curated by what criteria, toward whose ends, with what transparency?

When a job platform shows you a curated list of positions, and that curation is based on what you're statistically likely to click rather than what you're statistically likely to thrive in, the interests are misaligned and the curation is obscured. When a health app shows you content based on engagement optimization rather than clinical value, same problem. When a financial product shows you instruments ranked by commission rather than suitability, same problem with fiduciary implications.

The user who thinks they're making a free choice from an open set of options is often making a constrained choice from a pre-filtered set, without knowing either the constraints or the filtering criteria. This is the deepest version of the illusion of control, and it is the one that AI has made most pervasive.

Real Control and What It Looks Like

Enough diagnosis. What does genuine user control actually look like?

It is not forty-seven toggles. It is not a privacy policy. It is not a consent banner with a small grey "reject" option. It is not a "why did I see this?" button that explains nothing.

Real control has a few identifiable properties.

It is **proportional**: the effort required to exercise a preference should be roughly proportional to the significance of what's being controlled. One click to accept should mean one click to decline. The asymmetry between opting in and opting out is always a signal that something is wrong.

It is **consequential**: changing a setting should produce a visible, meaningful change in experience. If turning off ad personalization makes no discernible difference to the ads you see, the control is not real.

It is **legible**: users should be able to understand, in plain language, what they're agreeing to, what the alternatives are, and what the consequences of each choice are. Not legal language. Not technical language. The language a reasonably intelligent person who has not read your documentation would understand.

It is **reversible**: any choice should be undoable, and undoing it should be as easy as making it. The asymmetric cancel flow, the impossible unsubscribe, the data that can't be deleted — these are all reversibility failures that reveal where the real interests lie.

And perhaps most importantly, real control is **honest about its limits**. There are things about AI-powered systems that cannot currently be made fully controllable — the model's behavior, the training data, the upstream decisions made before the interface is built. Acknowledging these limits, rather than papering over them with performative controls, is itself a form of respect for the user.

The Power of the Default

We've come back, as this book keeps doing, to the default.

Because in the end, the most consequential design decision in any system involving choice is not the range of options offered. It's where you start.

Default settings are, for most users, permanent settings. The research on this is overwhelming and consistent. People accept defaults at extraordinary rates, across virtually every domain. Not because they don't care, but because they have limited attention and the default is presented as the expected, normal, recommended state.

This means that every default setting in an AI-powered product is a policy decision. A decision about what most users will experience, made without their input, by the people who built the system. And that decision is made in a context where the interests of the platform and the interests of the user are frequently in tension.

The ethical design question is simple to state and hard to implement: **set defaults for the user's benefit, not the platform's.**

Default to less data collection, not more. Default to lower automation, not higher. Default to opt-in for sharing, not opt-out. Default to conservative notification frequency, not maximum engagement.

These defaults would cost the platforms money. There's no getting around that. And that is exactly why they're not the defaults, and exactly why they should be.

Chapter 9: Designing for the Human in the Loop

In which we finally stop diagnosing the patient and start talking about the cure, discover that "human-centered AI" is not just a conference theme, and learn that the best technology is the kind that makes you feel more like yourself, not less.

In 1968, a computer scientist named Douglas Engelbart gave a demonstration in San Francisco that people who were present still talk about with something approaching reverence. In ninety minutes, Engelbart and his team demonstrated, for the first time in public, a computer mouse, video conferencing, collaborative real-time document editing, and hypertext linking between documents. It became known as the "Mother of All Demos".

The computing world in 1968 was dominated by the idea that computers were tools for automating tasks — replacing human labor with machine efficiency. Engelbart had a different idea. He called it **augmentation**.

His thesis, developed over years of research at Stanford Research Institute, was that the goal of computing was not to replace human intelligence but to extend it. To make humans more capable than they could be alone. To use machines to amplify what was distinctly human — judgment, creativity, synthesis, wisdom — rather than to substitute for it.

More than fifty years later, we are still, mostly, not building what Engelbart had in mind.

We are building automation. We are building engagement optimization. We are building systems that replace human judgment rather than extending it, that capture attention rather than expanding capability, that treat the human as a variable to be managed rather than a person to be served.

Engelbart's augmentation vision is still waiting to be properly built. And it is, I would argue, the most important design brief in the technology industry right now.

What Augmentation Actually Means

The word gets used a lot. "AI-augmented workflows." "Human-augmented intelligence." "Augmented decision-making." It has become one of those terms that means whatever the person using it needs it to mean, which usually means it means nothing.

So let's be specific.

Augmentation, in Engelbart's sense, means a system that makes you more capable in ways that compound. Not just faster at a specific task today, but genuinely better — more informed, more skilled, more able to handle complexity — over time. The augmented human should be more capable than the un-augmented human, and that capability should belong to the human, not the tool.

The test is the one we introduced in Chapter 7: if the tool disappeared tomorrow, would you be better or worse at this than before you started using it? Augmentation passes this test. Automation fails it.

A tool that teaches you while it helps you is augmentation. A tool that helps you while quietly eroding the underlying skill is automation wearing augmentation's clothes. The distinction is not always obvious in the short term. It becomes very obvious in the long term, when the tool changes or disappears and you discover what you've actually been left with.

The Billion-Dollar Meetings

Let me do some math. It will be approximate. Bear with me.

Take the Fortune 500. Each company, let's say, has around ten thousand employees. A conservative estimate puts meetings at fifty percent of the average knowledge worker's week. The average fully-loaded cost of an employee — salary, benefits, overhead — runs to around sixty dollars an hour. Meetings typically last an hour.

Run those numbers across a forty-five week working year and you arrive at somewhere in the region of **two hundred and seventy billion dollars** spent annually, across the Fortune 500 alone, on people sitting in rooms together.

Now apply the filter that anyone who has worked in a corporate environment will recognize immediately: roughly half the content of the average meeting has genuine value. The other half is catch-up that could have been an email, status theater for managers who need to feel informed, agenda drift, the guy who uses every meeting as an opportunity to relitigate a decision made six months ago, and twenty-minute discussions about whether there should be a space on either side of a slash.

Import/export or *import / export*. I have been in that meeting. Twelve people attended. A follow-up meeting was scheduled to discuss implementation.

Which brings the estimated waste to somewhere north of **a hundred and thirty-five billion dollars** per year, conservatively calculated and almost certainly understated.

I hope someone with a stronger grip on corporate economics can sharpen that number. But even if I'm wrong by an order of magnitude, it remains a number that should make any organization that considers itself serious about efficiency stop and sit quietly with its thoughts for a while.

The Woman Who Never Said a Word

The most impressive meeting system I have ever encountered was not built by a technology company. It was not an AI tool. It was not a productivity platform with a freemium tier and a conference circuit presence.

It was a woman at a small non-profit publishing house, working on a budget so tight that every hour had to justify its existence.

She had no job title that I can remember, but her function was this: she made meetings work.

A meeting at this organization was planned only when a clear agenda had been agreed upon in advance. She set the time, booked the room, identified exactly who needed to be present — not who was on the relevant group email, not who might theoretically have a stake, but who was *required* — and wrote out the agenda points before a single invitation was sent. Because she had to be budgeted for, meetings were scheduled judiciously. The act of planning a meeting required justifying it.

On the day, she arrived early. She had chased out anyone lingering from a previous booking. The projector was running. On screen were two documents: one with the agenda, the time, the date, and the names of all invited attendees, each marked absent. As people arrived, she marked them present. Minutes from any relevant previous meeting were already pulled up for reference.

As the meeting progressed, she took notes in real time — not a verbatim transcript but a live edit, bulleting points, moving comments around as the discussion developed, organizing for logic and clarity as it happened rather than afterward. When someone made a point that mattered, she noted who had made it. When someone said something that didn't pertain to the agenda, she moved it to the second document, to be reviewed separately, without interrupting the flow.

There was a man in these meetings — there is always this man — who frequently said things that didn't quite make sense and had a tendency to wander off into personal grievances and tangential observations. She never wrote down a single word he said.

He never noticed.

I loved her for that.

By the time I had walked back to my desk after the meeting, there was an email in my inbox with two attachments. Clean notes. Clear action points. Everything that mattered, nothing that didn't.

In the year I worked there, I never heard her speak in a meeting. Not once.

That, as I have said to many people since, is how you do that.

How Corporate America Does It Instead

In the decade I have spent working in corporate America, I have attended hundreds of meetings that were the precise opposite of everything described above.

Meetings scheduled without agenda items, because scheduling a meeting is easier than deciding what it's for. Meetings where note-taking is either absent entirely or performed reluctantly by whoever was last to find a chair, producing a document that nobody reads and nobody acts on. Meetings where twenty people are invited because they happen to be on a group email, and perhaps four of them have any reason to be in the room, and perhaps two of those four will say anything, and one of those two will spend fifteen minutes on something that has nothing to do with why the meeting was called.

Meetings, in short, as a performance of busyness rather than an instrument of work.

This is not a mystery. Projects have dead space — periods where work is waiting on delivery, approval, or someone else's input, and the honest answer is that there is nothing productive to do for a few hours. Meetings fill dead space. They make people look busy. Contractors on hourly rates have a specific and understandable affection for them. Managers use them to assert presence and gather information in ways that could, in a better-organized world, be achieved without stopping twelve people from doing their actual jobs.

The result is the hundred-and-thirty-five-billion-dollar waste. Sustained, year after year, by organizations that consider themselves efficient.

What AI Could Actually Do Here

Here is where this stops being a complaint about corporate culture and starts being relevant to everything this book has been arguing.

The meeting secretary at the non-profit was, in every structural sense, a model of good augmentation. She handled the parts of the meeting that didn't require judgment — logistics, scheduling, attendance, real-time note organization — so that the people in the room could focus entirely on the parts that did. She made the handoff clean. She had a clear scope. She knew what to write down and what to leave out. She produced output that was immediately useful without requiring further processing.

She was, in the language of Chapter 9, designing for complementarity. The humans did the thinking. She did everything else, better than the humans could have done it themselves.

This is precisely the brief that AI meeting tools are being designed to fulfill. Automatic transcription. Real-time summarization. Action point extraction. Follow-up scheduling. The technology exists, it is improving rapidly, and in its best current implementations it is genuinely useful.

But here is where the design question bites.

The meeting secretary's most important skill was not transcription. It was **judgment about what mattered**. She knew which comment was worth recording and which was noise. She knew which action point was real and which was something someone said to sound decisive. She knew — silently, without drawing attention to it — that the man who ranted was not contributing anything the notes needed to carry.

An AI transcription tool records everything. It does not yet know the difference between the point that will shape the project and the point that everyone in the room has already mentally discarded. Its summary is a compression of what was said, not a distillation of what mattered. And when it extracts action points, it extracts them from the transcript — which means it captures the things people said they would do, not necessarily the things that need doing.

This is not an argument against AI meeting tools. It is an argument for designing them with an honest model of what the human is still better at, and building the interface accordingly. The AI handles the transcript. The human — or ideally, a human who has been freed from the burden of transcription by the AI — handles the judgment layer. The action points get a review pass. The summary gets a read before it goes out.

The meeting secretary and the AI are not in competition. Used well, they are the same system, with different components handling different parts of the job.

Used badly — which is to say, used in a way that treats the AI summary as final, the extracted action points as authoritative, and the human as optional — they produce something that looks like efficiency and functions like the hundred-and-thirty-five-billion-dollar problem, just with better formatting.

Three Simple Rules

I have, over the years, developed a proposed taxonomy for meetings that I offer here free of charge to any organization willing to use it.

A **strategic meeting** justifies one hour. It is about decisions with significant consequences, requires the people authorized to make those decisions, and produces a clear record of what was decided and why.

A **tactical meeting** justifies thirty minutes. It is about coordination, status, and near-term problem-solving. It requires only the people directly involved in the relevant work.

A **stand-up** justifies fifteen minutes. It is about immediate blockers and daily priorities. It requires only the people whose work is blocked or blocking.

If you do not know, before scheduling a meeting, which of these three types it is — if you cannot answer that question in one sentence — there should not be a meeting. There should be an email, or a document, or a conversation between two people, or nothing at all.

This taxonomy is not sophisticated. It does not require a productivity platform or an AI tool or a workshop to implement. It requires only that the person scheduling the meeting ask, before sending the invitation: *what is this for, and who actually needs to be here?*

Two questions. Asked honestly. Applied consistently.

The saving, at scale, is left as an exercise for the reader.

The Complementarity Principle

There is a term I want to put at the center of this chapter: **complementarity**.

Complementarity means designing so that the human and the AI together outperform either alone — not by having the AI do more and the human do less, but by having each do what it's genuinely better at, in a way that covers the other's weaknesses.

What AI is genuinely better at: processing large amounts of data quickly, finding patterns across more examples than any human could review, maintaining consistency across repetitive tasks, being available at three in the morning, not getting tired or emotionally triggered.

What humans are genuinely better at: setting goals, applying contextual judgment that draws on lived experience, navigating ethical complexity, recognizing when the rules don't apply, caring about outcomes in a way that isn't reducible to a metric, knowing what matters.

Complementary design puts AI on the first list and humans on the second, deliberately and structurally. It does not let the boundary drift — AI gradually absorbing more of the second list because it's cheaper and faster and the quarterly numbers look better. It holds the line, because the second list is where the value that can't be automated lives.

In practice, this means designing handoffs — the seams we discussed in Chapter 2 — so that what passes from AI to human is genuinely the part that requires human judgment, not just the part the AI hasn't gotten around to yet. It means designing feedback loops so that human corrections actually improve the system rather than disappearing into a void. It means designing incentives so that users who engage critically with AI outputs are rewarded rather than penalized for the extra effort.

None of this happens by default. All of it requires deliberate design.

Passport to Success

In 2017, the United States Digital Service sent a team to fix the American passport renewal process. Not a large team. Not a multi-year enterprise transformation program with a steering committee and a governance framework and a budget that required congressional approval.

Six people.

A product manager, three engineers, someone whose job title was officially "bureaucracy hacker" — a role description that deserves wider adoption — and a UX designer. They set up in offices directly above the Washington DC Passport Agency, which turned out to be the most important decision they made. Not the methodology. Not the technology stack. The location.

Because being directly above the passport agency meant they could walk downstairs and watch real people, in real time, struggling with the real process they were trying to fix. They could observe where the form became unclear. Where people hesitated. Where confidence collapsed and the applicant started second-guessing information they almost certainly knew. They could talk to people waiting for their passports — not in a research lab, not in a scheduled usability session, but in the place where the frustration was actually happening.

This is what the United States Digital Service calls designing with users, not for them. The distinction is not semantic. Designing for users means constructing a model of what users need and building to that model. Designing with users means putting yourself in the room with the actual humans and letting what you observe change what you build. The first approach is faster. The second approach is more likely to produce something that works.

What they observed shaped every significant design decision that followed. They put qualification information at the beginning of the flow — if you didn't meet the criteria for online renewal, you found out in the first thirty seconds rather than after completing a lengthy form. They asked only the questions the process actually required, with additional fields appearing only when genuinely needed rather than presenting every possible field to every possible applicant regardless of relevance. They tested on paper before writing code — printing screens, putting them in front of real applicants, watching what happened, changing what didn't work, and repeating until the paper version worked before the digital version existed.

The results, when the system eventually launched, were not marginal. Processing times dropped to roughly one third of what they had been the previous year. When the pilot opened in August 2022, twenty-five thousand applications came in within the first two weeks. By 2024 the system was fully operational, with an estimated five million Americans eligible to use it annually.

None of this required artificial intelligence. None of it required machine learning or predictive algorithms or behavioral profiling. It required six people, a good location, the willingness to watch rather than assume, and the discipline to test on paper before spending money on code.

This is what the Dumbest in the Room looks like at government scale. A small team that positioned itself to be continuously corrected by the actual users of the actual system, and built accordingly. The technology served the process. The process served the human. In that order.

The bureaucracy hacker, for what it's worth, remains one of the best job titles in the history of government service.

Passport to Failure

Around the same time the Americans were fixing their passport process, the British government was attempting something similar — and running into a different problem entirely. One that is, if anything, more instructive.

The UK passport service had been assessed by the Government Digital Service against its Digital by Default Service Standard. The assessment panel, by all accounts, was impressed. The team knew what they were doing. The user research was solid. The frontend design — the interface that applicants would actually see and interact with — was well-considered and tested. The panel approved the service to proceed to beta.

But buried in the assessment report was a finding that deserves more attention than it typically receives in case studies of government digital transformation.

The service manager, the report noted, had effective power over the digital frontend. What they did not have effective power over was the backend — the processing systems, the databases, the infrastructure that sat behind the interface and actually handled the application once submitted. The backend was being managed under a separate project, constrained by an existing contract and change control processes that limited what could be altered. The frontend team and the backend team were, in the most consequential sense, not the same team.

The implication was clear and the assessment named it directly: the service manager might be limited in their ability to make the service genuinely digital end-to-end, because the system behind the front door hadn't fundamentally changed.

This is the seam problem made structural.

You can design a front door that is clear, welcoming, and efficient. You can reduce friction at the point of entry to near zero. You can make the first interaction so smooth that users feel, for the first time, that a government service has been designed for them rather than at them. And then the application disappears into a backend built on decades-old architecture, processed by systems that were never designed for the digital workflow the new frontend implies, managed by a team operating under a separate contract with separate priorities.

The user, of course, sees none of this. They see a good form followed by a wait. The wait is the backend. The backend is outside the control of the people who made the form good.

This is a pattern so common in enterprise digital transformation that it has its own informal name among practitioners: **the lipstick problem**. A well-designed interface applied to an unreformed underlying system. The experience looks modern. The process is not. And the gap between what the interface promises and what the system delivers is where user trust goes to die — slowly, in the form of unexplained delays, inconsistent responses, and the quiet realization that the impressive front end was essentially decorative.

The lesson is not that frontend design doesn't matter. It does. The lesson is that the seam — the handoff between what the user sees and what happens next — is as much an organizational design problem as a UX design problem. Fixing the interface without fixing the process behind it is not transformation. It is renovation. And renovation, however well executed, will eventually reveal the building it's been applied to.

The Americans put their team above the passport agency and watched users struggle with the real system. That proximity forced them to confront the whole problem, not just the visible part.

The British assessment identified a team that had done excellent work on the part they controlled, and were constrained by the part they didn't.

Both stories are true. Both are instructive. Good design requires access to the whole system, not just the front door. And the most important conversation in any digital transformation project is often not about the interface at all. It's about who controls the backend, under what contract, with what ability to change.

The bureaucracy, as it turns out, is part of the UX.

Progressive Disclosure of Automation

One of the principles I keep returning to in my own practice — and one that I think is underused in the industry — is what is called **progressive disclosure of automation**.

You may know progressive disclosure as a standard UX principle: don't show users everything at once. Surface the most important information first, and let users go deeper as their needs require. It's a principle about managing complexity by revealing it gradually rather than all at once.

Applied to automation, the principle works like this: **start users with more control and more visibility, and introduce automation incrementally as trust is established and capability is demonstrated.**

Don't start users on full autopilot. Start them doing the task, with AI assistance that is visible and optional. Let them see what the AI is doing and why. Let them build a calibrated sense of when the AI is reliable and when it isn't. Then, as that trust is earned through demonstrated accuracy in specific contexts, allow the automation to increase — with the user actively choosing to extend it, rather than having it imposed by default.

This is the opposite of how most AI products are currently designed. Most AI products default to maximum automation and provide controls that allow users to reduce it — controls that, as we established in Chapter 8, most users never touch. Progressive disclosure of automation inverts this: default to minimum automation, provide a clear pathway to increase it, and treat that increase as a deliberate act by an informed user rather than an inert acceptance of a default.

The short-term experience of this approach is worse. Users who are defaulted to full automation have a smoother initial experience than users who are asked to engage more. The long-term experience — capability maintained, trust calibrated, dependency avoided — is significantly better. This is, again, a values question disguised as a design question. Which outcome are you optimizing for?

Designing the Override

If there is one concrete design principle that I would require every AI product team to implement before shipping, it is this: **the override must be as easy as the accept.**

If a user can accept an AI suggestion in one click, they must be able to reject or modify it in one click. If an AI-generated output can be sent in one action, it must be correctable in one action. If automation can be activated in a single setting, it must be de-activatable in a single setting.

This sounds obvious. It is almost universally violated.

The asymmetry between accepting and overriding AI outputs is one of the most consistent dark patterns in enterprise AI design. Accept flows are smooth, prominent, and require minimal cognitive effort. Override flows require hunting for options, navigating warnings, and sometimes explaining yourself to a system that responds with reluctance — as if overriding its output is an unusual and faintly suspicious act.

It is not an unusual act. It is the central act of a human-in-the-loop system. And designing it as an afterthought or an obstacle is a structural statement that the system does not actually trust users to override it — which is, one way or another, always a lie.

Good override design is not just usable. It is **welcoming**. It treats correction as valuable — which it is, both for the immediate task and for improving the system over time. It makes the path from "I disagree with this output" to "I have changed this output" direct, fast, and consequence-free. It does not ask the user to justify themselves. It does not add friction to discourage correction. It makes correction the easy thing, because correction is the right thing.

The Feedback Loop That Actually Closes

Closely related to override design is the question of what happens to the override after you make it.

In most AI systems: nothing visible.

You correct an output. The output is corrected. You move on. The system produces the same kind of output next time, having apparently learned nothing from your correction, and you correct it again, and again, in what becomes a Sisyphean routine that gradually convinces you the system is incapable of improvement and you should probably stop bothering.

This is a closed feedback loop in name only. The correction is logged somewhere. Maybe it influences a future model update. Maybe it doesn't. You have no way to know, because the loop never closes visibly.

Genuinely closed feedback loops have three properties: the correction is **acknowledged** (the system responds to it in some way), it is **applied** (the behavior changes in the current context, immediately if possible), and it is **retained** (the system is observably better at this type of task in future interactions).

When all three are present, something remarkable happens. Users stop thinking of the AI as a black box that produces outputs of unpredictable quality and start thinking of it as a collaborator that improves with engagement. The relationship changes. Trust becomes calibrated rather than brittle. Users invest in the system because the system demonstrably rewards investment.

This is not technically difficult to design. It is organizationally difficult, because closing the feedback loop visibly requires coordinating across product, engineering, and model teams in ways that are slower and more complex than just logging the correction and moving on. The path of least resistance is the unclosed loop. But the path of least resistance is also the path to systems that users abandon as soon as a better option appears.

Graceful Degradation

Every AI system will fail. Will be slow. Will be unavailable. Will produce outputs so wrong that no amount of correction will fix them. Will encounter an input so outside its training distribution that it returns something that makes no sense at all.

The question is not whether these failures will happen. The question is what the user experiences when they do.

Graceful degradation means designing the failure modes as carefully as the success modes — ensuring that when the AI can't help, the experience falls back to something usable rather than collapsing entirely.

This is partly a technical problem. But it's largely a design problem. Because graceful degradation requires that the system knows and communicates its own limitations — which, as we discussed in Chapter 6, requires a kind of designed humility that most AI products don't currently practice.

A graceful degradation looks like: "I'm not confident about this — here's what I found, but you may want to verify before acting." Or: "I don't have enough information to give you a reliable answer here — can you add more context?" Or simply: "This type of request is outside what I handle reliably — here's a path to a human alternative."

Each of these is a system acknowledging a limit. Each of them is also, counterintuitively, a trust-building moment. Users who discover a system's limits from the system itself — clearly, honestly, early — develop calibrated trust: they know what to rely on and what to check. Users who discover a system's limits by acting on bad output develop algorithm aversion. The graceful failure is not a weakness. It is, in the long run, a feature.

Designing for Expert Users

Most AI UX is designed for the novice — the user who needs guidance, scaffolding, and smooth defaults. This is understandable. Novices are more numerous, their needs are easier to research, and designing for the novice produces better first impressions which drive adoption metrics.

But expertise matters, and AI products that don't design for expert users create a specific kind of problem: they make experts feel patronized.

Novices need scaffolding, examples, and safer defaults. Experts need something different: batch operations, shortcut pathways, granular controls, and — most importantly — a system that respects their competence rather than second-guessing it at every step.

An expert radiologist using an AI diagnostic tool does not need the system to explain what a lesion is. An expert lawyer using an AI research tool does not need the system to explain what a precedent is. An expert designer using an AI generation tool does not need a lengthy onboarding tutorial every time they open the application. These users have built their expertise over years. A system that treats them as beginners is a system they will abandon or work around.

Designing for expert users means building **progressive trust** into the interaction model. The system should be able to recognize, over time, that a user knows what they're doing and adjust accordingly — reducing explanatory scaffolding, increasing available depth and control, treating overrides as deliberate professional judgments rather than correctable errors.

This requires the system to model user expertise, which is technically feasible and organizationally rarely prioritized. It is worth prioritizing, because expert users are often the highest-stakes users — the ones making the most consequential decisions with the AI's outputs — and designing for their needs produces better outcomes than designing for the median novice and hoping experts adapt.

The Transparency That Actually Helps

We've talked at length about transparency as a trust-building tool. But there's a version of transparency that is worth focusing on specifically in the context of human-in-the-loop design: **operational transparency**.

Operational transparency means making visible, in real time, what the AI is doing and on what basis — not as an after-the-fact explanation but as a live part of the interaction.

This looks like the AI assistant that shows which documents it searched before producing a summary. The recommendation engine that surfaces the two or three factors most influencing a suggestion, not as a full explanation but as a quick-check: *based on your last three purchases and your location*. The diagnostic tool that highlights the specific features of an image it's attending to, so the expert can assess whether it's looking at the right things.

In each case, the transparency is not slowing the user down with unnecessary information. It's giving the expert user exactly what they need to make a rapid judgment about whether to trust this specific output. It's designed for a user who is actively engaged, not one who is passively accepting. It is transparency in the service of collaboration rather than transparency as compliance.

The design challenge is calibrating the depth. Too little transparency and the user can't check the system's work. Too much and the interface becomes a wall of metadata that nobody reads. The right amount is the amount that enables a competent user to form a rapid, calibrated judgment about this specific output in this specific context. Finding that amount requires research, testing, and the willingness to treat verification not as a cost but as a core user need.

When to Get Out of the Way

Here is the design principle that takes the longest to internalize, and that I find myself returning to most often in practice:

Sometimes the best AI UX is less AI.

There are tasks where human engagement with the full complexity of the problem — the struggle, the uncertainty, the process of working something out — is itself the point. Where the output matters less than what the process of producing it does to the person's capability, judgment, or understanding.

Designing AI into these tasks is not augmentation. It is substitution. And substitution in these contexts does not make users more capable. It makes them faster and more **fragile**, which is a trade that looks good in the short term and reveals itself as a bad deal later.

Knowing which tasks are which requires judgment that cannot be automated — which is perhaps the most elegant structural argument for why the human must remain in the loop. The system cannot know when its own use is counterproductive. Only the designer, thinking carefully about the long-term human implications of their choices, can make that call.

And making that call honestly — recommending less AI rather than more, in contexts where less is genuinely better for the human — is one of the hardest and most important things a designer in this era can do. It requires resisting the pressure toward feature completeness, toward impressive demos, toward the quarterly metric that rewards engagement regardless of its quality.

It requires designing, in the deepest sense, for the human on the other end of the screen.

A Manifesto in Four Sentences

I want to try to compress everything in this chapter — and, really, everything the last eight chapters have been building toward — into the smallest possible space.

AI should make you more capable, not more dependent. It should show its work, not hide it. It should make correction easy, not difficult. And when in doubt, it should do less and explain more.

That's it. Every design decision in an AI-powered product can be evaluated against those four sentences. Not every decision will satisfy all four simultaneously — there are real trade-offs, and pretending otherwise is its own kind of dishonesty. But they are the right questions to be asking, and they are questions that the industry is not yet, consistently, asking.

Chapter 10 is about why. And about what it would take to change that.

Interlude: Design Theory 47

I was nineteen years old, studying graphic design at the Gerrit Rietveld Academie in Amsterdam, and I was working on a poster.

This was the mid-1980s. Theater posters were having a moment. Album covers for LPs were still a legitimate art form — twelve inches of real estate that a designer could do something genuinely interesting with. The work felt important. It felt like it mattered what you put on the page, and more relevantly, it felt like *you* were the one putting it there. Your eye. Your instinct. Your particular way of seeing things.

I was deep in exactly that process — free-associating, following a thread of visual logic that only half-existed yet, jumping around in my head in the way that, to me, was the whole point of being a designer — when a visiting professor from Germany walked into the studio. He stopped behind me. He looked at what I was doing for a moment. Then he asked, in the way that visiting professors from Germany ask things, what design theory I was applying.

I told him I was free-associating.

He nodded slowly, in the way that visiting professors from Germany nod.

"Ah," he said. *"Design Theory 47."*

And he moved on.

My first reaction was something between deflation and mild outrage. Design Theory 47. He had taken the thing I considered most essentially mine — the random, intuitive, gloriously undisciplined way I moved through a design problem — and filed it. Categorized it. Given it a number. It was like being told that your treasured gift was an algorithm.

My second reaction, which took somewhat longer to arrive, was more interesting.

What if he was right? What if free association wasn't the absence of theory but a theory in itself — one approach among many, each with its own logic, its own appropriate contexts, its own strengths and failure modes? What if the other approaches I'd dismissed, a little arrogantly, as methods for designers who lacked the gift, were actually tools that more talented designers than me had developed because they needed them?

This is the question that a professional design career answers for you, slowly and sometimes painfully, over many years.

The answer, it turns out, is: both things are true, and the skill is knowing which one you need.

The volume and tempo of professional design work require process. You cannot free-associate your way through a six-week enterprise software project with fourteen stakeholders and a launch date. Consistency matters to clients in ways it doesn't matter to art students. Accountability — being able to explain, to the person paying for the work, why you made the choices you made — requires something more defensible than "it felt right on a Tuesday." Personal taste is subjective. Data is not. Research, metrics, A/B testing, user research, best practices — these are not the tools of designers who lack talent. They are the tools of designers who have learned that talent alone doesn't ship.

I committed fully to that professional apparatus. I use all of it. I have built a career on it.

And I have never stopped also doing Design Theory 47.

Not always for clients. Not always visibly. But somewhere in almost every project, I make the alternative version — the one based on gut feeling and a free association that I can't fully explain yet. Sometimes it goes nowhere. Sometimes it produces the thing that makes the whole project interesting, the thing that makes the client's product feel like itself rather than like everything else.

I keep doing it because it's why I became a designer. I wanted to jump around in my head. I wanted to follow the thread. I wanted, if I'm being completely honest, to have **fun** — and to make things that were worth having fun with.

This, I realize now, is not a luxury or an indulgence. It is load-bearing.

Because here is what Design Theory 47 actually is, beneath the joke: it is the part of design that cannot be systematized. The part that comes from having looked at thousands of things and developed an instinct that operates faster than conscious analysis. The part that knows, before the research confirms it, that this solution is more interesting than that one. The part that is, in the deepest sense, human.

Which brings me to AI.

As intelligent systems take on more of the systematic parts of design — the layout generation, the copy drafting, the A/B variant production, the pattern application — Design Theory 47 is not becoming less relevant. It is becoming more relevant. Because the systematic parts are exactly what AI does well, and the free-associating, gut-feeling, why-does-this-feel-right-to-me parts are exactly what it doesn't.

An AI can produce Design Theory 1 through 46 faster and cheaper than any human designer alive. It cannot produce Design Theory 47, because Design Theory 47 is not a theory. It is a person, following a thread that only they can follow, making something that only they could have made.

The designers who will matter most in the AI era are not the ones who resist the systematic tools. They are the ones who use those tools for what they're good at, and reserve something — some irreducible, unjustifiable, gloriously unnumbered instinct — for themselves.

The visiting professor gave it a number. He was telling me that what I was doing was real, and recognized, and had a place in the taxonomy.

But I think he knew, and I think I knew, that the number was a joke.

Some things resist categorization. In design, as in life, those are often the most important things.

Chapter 10: The Designer's Responsibility

In which we stop pointing at the industry and start pointing at ourselves, discover that neutrality is a myth, and make the argument that the most important skill a designer can have right now is the courage to say "this isn't good enough."

In 1945, a physicist named Joseph Rotblat did something almost nobody in his position had ever done before.

He quit the Manhattan Project.

He was the only scientist to voluntarily leave before the bomb was completed — not for personal reasons, not because the work was too hard, but because he had concluded that the original justification for building the bomb no longer applied. Germany was losing. The bomb, he decided, was no longer a defensive necessity. It was becoming something else. And he wanted no part of what it was becoming.

Rotblat spent the rest of his long life working on nuclear disarmament. He co-founded the Pugwash Conferences, a series of meetings between scientists and policy makers aimed at reducing the dangers of weapons of mass destruction. In 1995, he and Pugwash were awarded the Nobel Peace Prize.

I am not comparing UX designers to nuclear physicists. The stakes are not the same. Nobody has ever been vaporized by a bad onboarding flow. Not yet anyway.

But the structure of Rotblat's choice — the recognition that technical skill creates moral responsibility, that building something because you can is not the same as building something because you should, that at some point the professional and the ethical are not separable — is a structure that designers need to be having a serious conversation about right now.

We are not having it seriously enough.

The Myth of the Neutral Designer

The comfortable story that the design profession has told itself, for most of its history, goes something like this: designers are translators. We take a requirement — from a business, a client, a product manager — and we **translate** it into an experience. Our skill is a craft. Our job is the execution. The values embedded in the thing we're building are someone else's responsibility. We just make it usable.

But with the empowerment that AI is giving designers, I believe it is a form of professional negligence. Every design decision embeds a value. The choice of default. The prominence of a button. The language on a notification. The presence or absence of a friction point. The decision to include an override or to make it difficult. None of these are neutral. All of them shape behavior, at scale, for millions of people. And the person making those choices — the designer — is the one who has the most direct influence over their form.

Langdon Winner, a political theorist, argued in 1980 that artifacts have politics — that the things we build embed social choices, power relationships, and value systems, often invisibly. He was writing about bridges and city planning.

Looking at you, Robert Moses.

The argument applies with far greater force to AI-powered digital products, which have a reach, a behavioral influence, and an adaptability that no bridge or parkway has ever had.

The interface is not neutral. The algorithm is not neutral. The designer is not neutral. These are all active participants in shaping human experience and human behavior, at a scale that most designers have not fully reckoned with.

Reckoning with it is overdue.

The Decisions Nobody Sees

The decisions that matter most in design are usually not the ones that get presented to stakeholders.

The stakeholder sees the final design — the screens, the flows, the visual language. What they don't see are the hundreds of micro-decisions made in the process of getting there. The default setting. The copy on the decline button. The number of steps in the cancellation flow. The threshold at which the AI acts automatically versus asking for confirmation. The prominence of the privacy control relative to the accept button.

These decisions are made, in most organizations, by designers and product managers, often quickly, often under pressure, often with incomplete information about their downstream effects. They are rarely reviewed by ethics boards. They are rarely subjected to the kind of scrutiny that the big, visible decisions receive. And they are, collectively, the decisions that determine whether a product respects its users or exploits them.

This is where the designer's responsibility is most acute and most invisible. Not in the grand design choices that get debated in workshops, but in the small, fast, unwitnessed choices that shape the experience at the granular level where behavior actually lives.

The designer who sets a default knows they're making a policy decision for millions of users. The designer who writes a confirm-shaming button label knows, at some level, what they're doing. The designer who makes the override flow deliberately cumbersome knows the effect that will have.

Most of the time, they make these choices without naming them. Without saying, in the meeting: this default will mean that 90% of users share data they would not choose to share if the question were asked neutrally. Without saying: this button label is designed to make users feel bad about exercising a legitimate preference. Without saying: this cancellation flow is designed to exhaust users into staying.

The first act of professional responsibility is naming what you're doing.

The Room You Walk Into

Before we close this book with principles and manifestos and calls to arms, let me describe a room.

It is a kick-off meeting for a digital transformation project. You are the UX designer. You have been briefed, you have done some preliminary research, and you are genuinely excited. You have ideas. You have a process. You are, in the language of the industry, ready to be disruptive — which is a word that sounds exciting in a portfolio and means something considerably more uncomfortable in practice.

The corporation you are about to help is called, let's say, ABCD. It has 120,000 employees. Its internal enterprise software is used every day to keep the business running, and it is failing constantly. Employees have taken to using paper and pencil to capture data when the systems go down, which is often. A technology audit has revealed that the software currently in use comes from thirty different suppliers. Some of those suppliers no longer exist. Ten of them have upgraded their products and no longer support the versions ABCD is running. One machine — one machine that is load-bearing for actual business operations — still runs Windows 3.x. The programmer who wrote the critical software that runs on it died two years ago. Of old age.

Think the millennium bug, but distributed across an entire organization and left to quietly worsen for twenty years. Not a bug, but an entire ant hill.

Senior management has decided that something must be done. The solution, as it always is in these situations, is a portal. With a dashboard. Bespoke, proprietary, maintained in-house, doing everything. Not a bad idea in principle. Extraordinarily difficult in practice, because the thirty-supplier archaeology project required to understand what the portal needs to replace hasn't been done, can't easily be done, and nobody is entirely sure where to start.

The heat is further raised by the implicit threat that if the internal teams can't sort this out, management will bring in one of the large enterprise software firms — the 600-pound gorillas who will arrive, rationalize everything with cheerful brutality, and redistribute the organizational power that various fiefdom managers have spent careers accumulating.

So the stress levels, as you walk into the kick-off meeting, are considerable.

What you are about to discover is that no single person in this organization understands the business in its entirety.

Not the CTO. Not senior management. Not the BAs, PMs, POs, and PLs — all the esoteric titles that populate the meeting room, all of whom consider themselves better UX designers than you because they know their product, which means they know their corner of it, which is precisely how the organization ended up with thirty suppliers and a machine running Windows 3.x.

Your job, as the UX designer, is to create a holistic overview of the workflow. To map the whole thing. To ask the questions that produce a picture of how the organization actually operates rather than how it believes it operates.

This is perceived as an attack.

Not intentionally. You are not trying to make anyone look bad. You are trying to understand the system so you can design for the people who use it. But the act of asking — of making visible the gaps and inconsistencies and redundancies that nobody has formally acknowledged — makes abundantly clear that management has not been keeping an eye on the ball. And nobody in that room wants to be the person who was supposed to be keeping an eye on the ball.

They will want to kill the messenger. The messenger is you.

The project will proceed as follows.

You will try to do research. You will sketch, show designs, facilitate workshops, and attempt to brainstorm with people who have been in adjacent meetings for so long that the concept of genuine open inquiry feels threatening. You will make something that one BA asked for. A PM, in a different meeting you weren't invited to, will request changes. The BA will ask to change it back. The PM will not have told the BA about the meeting. You will be caught between them, producing revisions that cancel each other out, in a process that generates the appearance of progress while producing none.

You will become a pawn in office politics that were established long before you arrived and will continue long after you leave.

The project will go over budget. It will be delivered late. When it launches, the users will complain — because it is new, and users do not like new, and enterprise software users particularly do not like new because new means retraining, and nobody wants to retrain or be retrained, and the budget for training was one of the first things cut when the project went over budget.

There will be bugs. There are always bugs. Beta software in a 120,000-person organization surfaces failure modes that no testing environment could have anticipated.

And when the complaints come in — from users, from business owners, from the managers whose fiefdoms were disrupted — they will be directed, with the unerring accuracy of organizational self-protection, at the most visible external party in the process.

The UX designer.

I am not telling this story to be discouraging. I am telling it because it is true, and because the principles in this book — design for the human, hold the default to a high standard, name the dark pattern when you see it, ask who this design actually serves — are principles that need to survive contact with this room.

A room where the politics are real. Where the budgets are constrained. Where the timeline was unrealistic before you joined. Where the people who should be your allies are fighting each other. Where the user — the 120,000 employees who need the software to work so they can do their jobs — is the least powerful person in the conversation and the most affected by the outcome.

The ethical design practice described in this chapter is not built for ideal conditions. It is built for this room, because this room is where most design actually happens.

The principles don't get easier to apply when the room is difficult. They get more important.

That is precisely when they matter most.

The Courage Problem

Naming it is not enough, of course. Naming it in a room where the business case points the other way is where the courage comes in.

I want to be honest about how hard this is, because books about professional ethics have a tendency to make the right choice sound obvious and easy, and then practitioners read them and feel vaguely guilty for not having made the right choice more often, and nothing changes.

The right choice is not obvious. It is not easy. The A/B data usually points the other way. The business case is usually clear and present. The ethical cost is usually diffuse and delayed. The person in the meeting who says "I'm not comfortable with this" is the person who is slowing things down, raising problems without solutions, making their colleagues' jobs harder.

This is the structure of most ethical failures in organizational life: not a dramatic choice between good and evil, but a series of small choices between comfort and discomfort, between going along and pushing back, between "this is someone else's problem" and "this is my problem too."

The design profession has not built adequate institutional support for people who want to push back. There is no equivalent of the medical profession's ethical framework, no formal code that gives practitioners professional cover for refusing to do something harmful. There are ethics guidelines from professional bodies — the UXPA has a code of professional conduct, the ACM has principles — but they are largely aspirational documents that have no enforcement mechanism and no teeth.

What exists instead is individual courage, distributed unevenly, exercised inconsistently, and regularly punished by organizations whose incentives run the other way.

This is not a sustainable situation, and it is not good enough for a profession whose work now shapes the cognitive and emotional lives of billions of people.

What an Ethical Design Practice Looks Like

I am not going to pretend that I have always gotten this right. Thirty years in enterprise software design means thirty years of compromises, of decisions made under deadline pressure, of business cases that won arguments they should have lost. I have shipped things I wasn't proud of. I have made defaults that served the client rather than the user. I have, on occasion, been the person who named the problem and then let it go when the pushback was too strong.

What I have learned from that record — and from the experiences of designers I respect who have navigated this better than I have — is that ethical design practice is not a destination. It is a discipline. It requires active, ongoing maintenance, like a skill that atrophies if you don't practice it.

It looks like a few specific habits.

Ask the question nobody is asking. In every product discussion involving AI, someone should be asking: what happens to the user when this goes wrong? What are we optimizing for, and is that what we should be optimizing for? Who is not in this room whose interests are affected by what we're deciding? These questions are often uncomfortable. They are never irrelevant.

Name the dark pattern when you see it. Not after the meeting, in the corridor, to a sympathetic colleague. In the meeting. On the record. "This default is designed to make users share data they wouldn't choose to share if asked directly. I want to flag that before we finalize it." You may not win the argument. But you have named it, and naming changes things — slowly, incrementally, more than silence does.

Design the failure mode first. Before you design the happy path, design what happens when the AI is wrong. What happens when the user wants to override. What happens when trust has been broken. Systems that have well-designed failure modes are systems that were built by people who took the human seriously. This practice is also, pragmatically, the fastest way to identify the ethical weak points in a design — because failure modes are where the real values of a system are revealed.

Use the AI contract honestly. If you can't write, in one sentence, what your AI feature does for the user — inputs, outputs, action, constraints — it is not ready to ship. Not because the sentence is a bureaucratic requirement but because the inability to write it clearly means the design has not grappled honestly with what it's doing. The sentence forces clarity. Clarity makes the ethical dimensions visible.

Hold the default to a higher standard. For every default setting in an AI-powered product, ask: if a user knew exactly what this default meant for their data, their attention, and their capability over time, would they choose it? If the honest answer is "probably not," the default needs to change, or the decision to ship it anyway needs to be made consciously and explicitly by someone with the authority and the accountability to make it

The Specific Responsibilities of AI-Era Design

The general principles above have always applied to design. What's new in the AI era is a set of specific responsibilities that didn't exist, or didn't exist at this scale, before.

The responsibility of legibility. As AI systems become more complex, the gap between what users believe a system is doing and what it is actually doing grows. Closing that gap — making AI behavior legible to the humans it affects — is a design responsibility. Not a nice-to-have. Not a feature for the next release. A baseline obligation.

The responsibility of honesty about limits. AI systems have limits. They hallucinate, they drift, they fail in edge cases, they are confidently wrong in ways that traditional software is not. Designing these limits into the communication of the product — not hiding them behind smooth interfaces and optimistic copy — is a responsibility. Users who don't know the limits can't protect themselves from the failures. Designers who don't communicate the limits are not protecting users. They are protecting the product's short-term reputation at the user's long-term expense.

The responsibility of long-term human outcomes. We have spent this entire book making the case that AI-powered UX has consequences for human capability, human autonomy, and human wellbeing that play out over years, not sessions. Designing with those consequences in mind — asking not just "does this work?" but "what does this do to the person who uses it for five years?" — is a responsibility that the industry has not yet institutionalized but must.

The responsibility of the people who are not in the room. The Sophias on the sofas. AI systems affect people who are not users of those systems. The facial recognition database affects people who never consented to be in it. The algorithmic content feed affects the social fabric of communities whose members use it. The automated hiring system affects candidates who have no idea it exists. Designing for these affected parties — the people who are not the primary user but who are nonetheless subject to the system's decisions — is a responsibility that expands the designer's scope far beyond the screen.

Bias

Consider a retail recommendation engine designed to maximize sales of a specific product category. If that engine is trained on purchasing data that skews toward a particular demographic, it will systematically under-serve or exclude others — suggesting generic sizing that doesn't fit, or defaulting to options that reflect one customer profile while ignoring others entirely. Profits may look good in the short term. The long-term cost, in alienated customers and reputational damage, compounds quietly until it doesn't.

The Industry We Could Build

Here is where I want to end, because books about problems are only useful if they gesture, however imperfectly, toward something better.

The industry we have is not the industry we have to have.

We have, right now, more UX designers, more AI researchers, more product thinkers, and more computational capability than at any point in history. We have a profession — design — that has always defined itself, at its best, by advocacy for human experience. We have tools that, used honestly, could genuinely extend human capability in ways that Engelbart dreamed of in 1968 and that remain, mostly, unrealized.

The gap between the industry we have and the industry we could build is not primarily a technical gap. The technology is there, or close enough. The gap is a values gap. A gap between what we optimize for and what we should optimize for. Between short-term engagement and long-term capability. Between the metric on the dashboard and the human on the other side of the screen.

Closing that gap requires designers who are willing to name what they're doing, to push back when the business case and the ethical case diverge, to hold the default to the standard it deserves, and to ask — always, persistently, even when it's uncomfortable — whose interests this design actually serves.

It requires organizations that create space for those conversations rather than treating them as obstacles to shipping. It requires leadership that understands, at a genuine level, that user trust is a long-term asset that dark patterns erode faster than any feature can rebuild. It requires a profession that develops the institutional structures — the ethics frameworks, the professional standards, the shared language — to support practitioners who want to do the right thing and currently have to do it alone.

None of this is easy. All of it is possible.

The Title, Revisited

This book is called *Don't Make Me Think, But Don't Think For Me.* I want to come back to that title here at the end, because I think it contains everything that matters.

Don't make me think. Design for clarity, for ease, for the removal of unnecessary friction. Respect my time and my attention. Don't make me puzzle over things that shouldn't require puzzling. Steve Krug was right in 2000 and he's still right now.

But don't think for me. Don't replace my judgment with yours. Don't optimize my experience for your metrics while framing it as my benefit. Don't make me dependent on your system while calling it empowerment. Don't remove my agency while presenting me with forty-seven toggles and calling it control.

The space between those two instructions is not a narrow gap. It is the entire territory of good design. It is large enough to build a career in, complex enough to spend a lifetime navigating, and important enough — right now, in this moment of extraordinary AI capability meeting ordinary human psychology — to take with absolute seriousness.

The user on the other side of the screen is a person. They have a life outside your product. They have capabilities you should be making stronger, not weaker. They have interests that may not align with your engagement metrics. They have the right to understand what your system is doing, to correct it when it's wrong, and to leave it if they choose.

They are not a behavioral profile to be optimized. They are not a session to be extended. They are not a conversion to be maximized.

They are a person.

Design like it.

Epilogue: After the Screen

This book has been about UX and AI. About the interface as a visible, navigable, clickable thing. About the design decisions that shape what you see, what you can do, and what is done to you, on a screen.

That screen may not be the primary surface of human-computer interaction for much longer.

The trajectory is clear enough. Voice interfaces are already handling tasks that previously required navigation. Agentic AI systems — ones that act on your behalf rather than waiting for instruction — are moving from novelty to infrastructure. The dropdown menu, the modal dialog, the cookie consent banner with its artfully hidden reject button — all of these assume a user sitting in front of a screen, making choices, clicking things. What happens to UX design when there is no screen to click?

The optimistic answer is that the friction disappears. You state an intent, the system executes it, the task is done. No hunting for buttons. No buried settings. No confirm-shaming. No hidden costs revealed at the final step.

The realistic answer is that the problems don't disappear. They migrate.

An AI agent that acts before confirming has no cancel button to find — because there is no button at all. A voice system that mishears and executes has no visible error state. An agentic workflow that completes a task you didn't quite intend produces no audit trail you can easily examine. The seam problem, the trust problem, the transparency problem — everything this book has argued about — survives the death of the GUI. In some respects it gets considerably worse, because at least with a screen you could see what the system was doing.

The principles in this book were written for the interface era we're in. They will need translation for the interface era that's coming. But the translation is not complicated.

There is still a human on the other end of this interaction. They still have interests that may not align with the system's. They still deserve to understand what is being done on their behalf, to correct it when it's wrong, and to remain capable rather than dependent.

The screen was never the point. The human always was.

Augmented Knowledge

We have been calling this technology artificial intelligence for so long that the name has stopped feeling strange. It shouldn't. As this book has argued, and as Albert demonstrated in the opening pages, there is nothing intelligent about these systems in any meaningful sense of the word. No awareness, no understanding, no judgment, no curiosity. Extraordinary capability, yes. Intelligence, no.

The name matters because names shape expectations. Call something intelligent and people expect it to understand them, to know when it's wrong, to exercise judgment in ambiguous situations. When it fails to do these things — and it will — the failure feels like a betrayal rather than a technical limitation. The gap between the name and the reality is itself a design problem, one that the industry has shown no interest in fixing because "artificial intelligence" sells products and "statistical pattern-matching system" does not.

If I were naming it, I would call it augmented knowledge. Not intelligence — knowledge. The systems we have built are extraordinary repositories of human knowledge, compressed and made accessible in ways that were previously impossible. They augment what we know, what we can find, what we can do with information. That is genuinely remarkable and genuinely useful. It is not thinking. It is not understanding. It is not intelligence.

Augmented knowledge is honest about what the technology is and what it isn't. It sets appropriate expectations. It positions the human as the intelligent party — the one who asks the question, evaluates the answer, and decides what to do with it. Which is, and has always been, exactly where the intelligence needs to live.

The UX of Robotic AI – The Door Test

How to tell if a humanoid robot is ready to live in your home, or is it just an expensive toy?

Like many people interested in humanoid robots, I think there's far too much "Wizard of Oz" marketing around consumer robots. Great work is being done in non-humanoid robotics, with machines running tirelessly in "lights-out" factories. But humanoid robotics are a unique and complex thing. Watching Elon Musk "dance" with his robots is a pure Wizard of Oz spectacle. Robotics is full of demo theatrics. Seeing robots do backflips and run (pre-selected videos often shown when the robot didn't face-plant), is at best an omission of the truth.

Standing up and opening doors in *uncontrolled, varied home environments* is a much harder **generalization** problem than choreographed stunts in a known environment. However, the simple things in life, like standing up from a chair, or opening a door, are incredibly complex. To stand up from a chair requires a delicate balancing act of a multitude of factors — like the height and type of the chair (for example, armrests or swivel), the angle and weight of the body.

Consider whether you have a robot in your home. In your house, you have many doors. We as humans effortlessly go in and out of rooms through doors. But doors are complex things in themselves. Here is a list of possibilities to be considered when going through a door.

- **Is the door locked?**
- **Does the door open outward or inward?**

- **On which side of the door has the knob, left or right?**
- **What type of knob is it — twist, pull, push? Does the robot have the hand dexterity to do this?**
- **Is it open already, or ajar?**
- **Does the door open and close on its own, or need to be pushed?**
- **Is it heavy or light?**
- **Is it a sliding door, a double door, a garage door or even maybe a swinging door?**
- **Is there anyone coming through the other way?**
- **Do you close it after you?**
- **Is there a door threshold to trip over?**
- **Are there steps going up to or down from the door?**
- **Do you give the robot a set of keys?**
- **Will the robot know which key to use for which door?**

If you are to welcome a robot into your house, you don't want to spend all your time opening and closing doors for it. I tried to find a video of a robot opening a door, but I couldn't find any convincing home door-opening demos in unconstrained settings outside of highly staged lab demos.

So, like the Turing Test, I'm opening up the Door Test: a simple, everyday feat.

The Door Test: A humanoid robot should be able to approach an unfamiliar household door, infer how it works, open it safely, pass through, and close it appropriately without human assistance or per-door programming.
I'm not opening the door of my house to any robot until it can do it itself.

About the Author:

I started working as a UX designer in print, but as many designers transitioned into the digital world, UX became an essential part of my tool kit. Print is passive, naively we entered the interactive digital world ripe for the slaughter.

Fortunately, I was inspired by the wonderful book "**Don't Make Me Think: A Common Sense Approach to Web Usability**" by Steve Krug, to become not only a graphic designer, but a designer that made valuable, meaningful products. But, as stated in that other great book "**The Inmates Are Running the Asylum: Why High Tech Products Drive Us Crazy and How to Restore the Sanity**" by Alan Cooper, technology can quickly become the Lord instead of the Servant. And in this exciting time of AI, this is truer than ever.

Don't Make Me Think, But Don't Think For Me is a personal journey through the joys and horrors of living with intelligent technology. With thirty years of experience designing enterprise software for some of the world's largest organizations — and a career that took me from art school in Amsterdam to the corporate corridors of New York — I know where the bodies are buried. I've helped build these systems. I've watched them fail. And I have some things to say about both.
Part field guide, part manifesto, and part confession, this is the book about AI and design that doesn't require a computer science degree — just a smartphone, a healthy suspicion that the "Accept All" button is not your friend, and the nagging feeling that technology was supposed to be working for you.

I am also a novelist. My published fiction includes *Amsterdam in Ultramarine, 1939: The Chronicles of Rishaan Finch*, *Citizen Robot*, and *Ghosts at War*. I live in Pound Ridge, New York, with my wife Jaana, who has the patience of a saint and the good judgment to tell me when I am wrong.
This is the first book I have written where the subject matter kept changing while I was writing it. It will not be the last.

ENDNOTES & REFERENCES

References are organized by chapter and keyed to the text by a short identifying phrase. Personal anecdotes and the author's own professional observations require no citation. All web sources were verified as accessible at the time of writing (2025). Where a study appears in multiple chapters, the full citation is given at first occurrence and subsequent references note where the full citation appears.

Before We Begin: The Dumbest in the Room

1. **"Don't Make Me Think"** — Krug, Steve. *Don't Make Me Think: A Common Sense Approach to Web Usability.* New Riders, 2000. Third edition published 2014. The title of this book and its central thesis are referenced throughout the text.

2. **"The Inmates Are Running the Asylum"** — Cooper, Alan. *The Inmates Are Running the Asylum: Why High-Tech Products Drive Us Crazy and How to Restore the Sanity.* Sams Publishing, 1999.

3. **Albert and large language models** — The 'Albert' metaphor for LLM mechanics and the frozen-time framing are the author's own formulation. For technical grounding on how language models work, see: Brown, Tom B., et al. 'Language Models Are Few-Shot Learners.' *Advances in Neural Information Processing Systems* 33 (2020). The 'compression' framing is attributed to Sam Altman in various public statements and interviews, 2023–2024.

Chapter 1: The Promise and the Problem

4. **Krug's thesis** — See note 1. The specific formulation 'don't make me think' and its implications for intuitive design are discussed throughout Krug (2000).

5. **The Nest Learning Thermostat** — Introduced in October 2011. For the pattern of user overrides being interpreted as new training data, see documented user reports and product reviews 2012–2015. Nest's public documentation describes the learning mechanism. The specific failure mode described — the thermostat 'learning the wrong lesson' — is widely reported in consumer reviews and technology journalism of the period.

Chapter 2: The Goldilocks Zone of Automation

6. **Eleanor Mure and the Goldilocks story** — Mure, Eleanor. *The Story of the Three Bears* (manuscript, 1831; privately printed c.1837). The story was subsequently retold by Robert Southey (1837) and the girl renamed 'Goldilocks' in later Victorian editions. The 1837 date in the text refers to the earliest printed version.

7. **Tesla Autopilot and the handoff problem** — The two-second handoff failure mode has been documented extensively in NTSB investigation reports of Autopilot-involved incidents. See: National Transportation Safety Board. *Collision Between a Car Operating With Automated Vehicle Control Systems and a Tractor-Semitrailer Truck Near Williston, Florida, May 7, 2016.* NTSB/HAR-17/02, 2017. The general pattern of driver disengagement under sustained automation is well established in human factors literature; see also Parasuraman, R., Sheridan, T. B., and Wickens, C. D. 'A Model for Types and Levels of Human Interaction with Automation.' *IEEE Transactions on Systems, Man, and Cybernetics* 30, no. 3 (2000): 286–97.

8. **The AI contract** — The AI contract framework (inputs → outputs → user action → constraints) is the author's own synthesis from enterprise AI UX practice. The conceptual foundations draw on Shneiderman, Ben. *Human-Centered AI.* Oxford University Press, 2022, and on the author's coursework and certification in AI and UX at Stanford University (2024–2025).

9. **US Digital Service passport project** — Nguyen, Annie. 'Moving the U.S. Passport Renewal Experience Online.' *United States Digital Service / Medium*, October 16, 2018. https://medium.com/the-u-s-digital-service/moving-the-u-s-passport-renewal-experience-online-a038e5d174ba. Processing time improvements cited from State Department announcements, September 2024.

10. **UK passport service assessment** — 'Passports — Service Assessment.' *Data in Government blog, UK Government Digital Service*. https://dataingovernment.blog.gov.uk/passports-service-assessment/. The finding about the service manager's limited control over backend systems is quoted directly from this published assessment.

Chapter 3: When the Machine Gets It Wrong

11. **Robert Julian-Borchak Williams** — Hill, Kashmir. 'Wrongfully Accused by an Algorithm.' *The New York Times*, June 24, 2020. Williams v. City of Detroit, filed 2021. The case is widely reported and was cited in congressional testimony on facial recognition. Williams's exact words at the interrogation — 'I hope you don't think all Black men look alike' — are reported in multiple accounts including Hill (2020).

12. **Résumé screener bias against women's colleges** — Amazon's internal AI recruiting tool is the most documented case. See: Dastin, Jeffrey. 'Amazon Scraps Secret AI Recruiting Tool That Showed Bias Against Women.' *Reuters*, October 10, 2018. The tool downgraded CVs that included the word 'women's' and graduates of all-women's colleges.

13. **Steven Schwartz / ChatGPT legal brief** — *Mata v. Avianca, Inc.*, Case 1:22-cv-01461 (S.D.N.Y. 2023). Judge P. Kevin Castel's sanctions order, issued June 22, 2023, is publicly available. Widely reported; see Weiser, Benjamin. 'Here's What Happens When Your Lawyer Uses ChatGPT.' *The New York Times*, May 27, 2023.

14. **Loss aversion / losses feel twice as significant** — Kahneman, Daniel, and Amos Tversky. 'Prospect Theory: An Analysis of Decision under Risk.' *Econometrica* 47, no. 2 (1979): 263–91. The approximate 2:1 ratio of loss to gain sensitivity is a finding of prospect theory and is well replicated.

15. **Algorithm aversion** — Dietvorst, Berkeley J., Joseph P. Simmons, and Cade Massey. 'Algorithm Aversion: People Erroneously Avoid Algorithms After Seeing Them Err.' *Journal of Experimental Psychology: General* 144, no. 1 (2015): 114–26. https://doi.org/10.1037/xge0000033. The finding is that participants who saw an algorithm perform—even outperform a human—were significantly less likely to choose it after observing a single error.

Interlude: Rembrandt's Beard

16. **DALL-E and AI image generation** — The author's own DALL-E experiments are documented with original prompts. The illustrated images reproduced in the book are the author's own AI-generated outputs. The observation about AI hand-generation errors is widely documented; for technical explanation see the literature on diffusion model limitations in compositional generation.

17. **Rembrandt's self-portraits** — Rembrandt van Rijn produced approximately 40–50 self-portraits over his career, providing an unusually comprehensive photographic-like record of his appearance. For reference: White, Christopher, and Quentin Buvelot, eds. *Rembrandt by Himself.* National Gallery Publications / Mauritshuis, 1999.

Chapter 4: The Disappearing User

18. **Netflix Prize** — The Netflix Prize ran from 2006 to 2009, offering $1 million for a 10% improvement in recommendation accuracy. The winning team was BellKor's Pragmatic Chaos. For the deployment decision and the gap between ratings-prediction and actual watch behaviour, see Amatriain, Xavier, and Justin Basilico. 'Netflix Recommendations: Beyond the 5 Stars (Part 1).' *Netflix Tech Blog*, April 6, 2012.

19. **Filter bubble** — Pariser, Eli. *The Filter Bubble: What the Internet Is Hiding from You.* Penguin Press, 2011. The term 'filter bubble' was coined by Pariser. Note: subsequent research on the scale of the effect has produced mixed results; see Guess, Andrew M., et al. 'How Do Social Media Feed Algorithms Affect Attitudes and Behavior in an Election Campaign?' *Science* 381, no. 6656 (2023).

20. **Artifacts have politics** — Winner, Langdon. 'Do Artifacts Have Politics?' *Daedalus* 109, no. 1 (1980): 121–36. Winner uses Robert Moses's low bridges on Long Island parkways as a canonical example of infrastructure embedding social values.

21. **Spotify fatigue** — The phenomenon of algorithmic playlist fatigue is widely discussed in music journalism. For a representative account see: Hogan, Marc. 'Has Streaming Killed Music Discovery?' *Pitchfork*, March 2019. The author uses 'Spotify fatigue' as an informal term for a documented listener complaint; it is not a clinical or formal research term.

Chapter 5: Dark UX — Designed Against You

22. **Dark patterns taxonomy** — Brignull, Harry. 'Dark Patterns: Dirty Tricks Designers Use to Make People Do Things.' darkpatterns.org, 2010. Brignull coined the term and built the original taxonomy cataloguing roach motels, confirmshaming, misdirection, hidden costs, and trick questions. The site has since expanded and moved to deceptive.design.

23. **Amazon Prime / FTC lawsuit** — Federal Trade Commission v. Amazon.com, Inc., Case No. 2:23-cv-00932 (W.D. Wash., filed June 21, 2023). The FTC's complaint describes in detail the multi-step cancellation flows Amazon used. Amazon agreed to simplify the process as part of the settlement.

24. **Ticketmaster / congressional hearing 2022** — US Senate Judiciary Committee hearing on 'That's the Ticket: Promoting Competition and Protecting Consumers in Live Entertainment,' January 24, 2023. The $23 facility fee exchange referenced in the text occurred during this public hearing. Ticketmaster's fee structures are documented in consumer complaints and journalistic investigation; see Knopper, Steve. *Appetite for Self-Destruction* (revised edition) for broader context on the ticketing industry.

25. **Facebook Papers / Instagram teen girls research** — Horwitz, Jeff, and Deepa Seetharaman. 'Facebook Knows Instagram Is Toxic for Teen Girls, Company Documents Show.' *The Wall Street Journal*, September 14, 2021. The internal Facebook research documents, subsequently known as the Facebook Papers, were provided to Congress and the press by whistleblower Frances Haugen.

26. **GDPR and pre-ticked boxes** — Under GDPR (Regulation (EU) 2016/679), recital 32 specifies that consent should be demonstrated by 'a clear affirmative act' and that pre-ticked boxes do not constitute valid consent. The UK ICO guidance is consistent with this interpretation.

27. **Cookie banner consent rates** — Nouwens, Midas, et al. 'Dark Patterns After the GDPR: Scraping Consent Pop-Ups and Demonstrating Their Influence.' *CHI Conference on Human Factors in Computing Systems*, 2020. The specific figure of up to 90% consent rates for dark-pattern banners versus 30–40% for neutral designs is drawn from this and related studies in the cookie consent literature.

Chapter 6: Trust, Transparency, and the Black Box

28. **Houston Federation of Teachers / EVAAS** — *Houston Federation of Teachers v. Houston Independent School District*, United States Court of Appeals for the Fifth Circuit, 2017. The case established that teachers had a due process right to meaningful explanation of the algorithm's outputs. The EVAAS system (Education Value-Added Assessment System) is a product of SAS Institute. For detailed analysis see: Kamenetz, Anya. 'The Algorithm That's Deciding Teachers' Fates.' NPR Ed, December 2016.

29. **The Sabrina Demming note** — The name 'Sabrina Demming' was generated by an AI chatbot when the author asked for a personalised account of the Houston teachers case. The name could not be verified as a real person. This is documented in the text as an example of AI hallucination — the system produced a plausible-sounding specific narrative around a real general situation.

30. **EU AI Act** — Regulation (EU) 2024/1689 of the European Parliament and of the Council (the AI Act), published in the Official Journal of the European Union, July 12, 2024. High-risk AI system requirements are set out in Articles 8–15 and Annex III.

31. **'Evidence beats explanation'** — This formulation reflects a principle that has emerged across AI interpretability and XAI (explainable AI) research. For academic grounding see: Doshi-Velez, Finale, and Been Kim. 'Towards a Rigorous Science of Interpretable Machine Learning.' arXiv:1702.08608, 2017; and Lipton, Zachary C. 'The Mythos of Model Interpretability.' *Queue* 16, no. 3 (2018): 31:31–57.

32. **Trust as competence, predictability, and alignment** — This three-dimensional model of trust in automated systems draws on: Lee, John D., and Katrina A. See. 'Trust in Automation: Designing for Appropriate Reliance.' *Human Factors* 46, no. 1 (2004): 50–80.

Chapter 7: Anxiety, Dependency, and the Atrophied User

33. **London taxi drivers and hippocampal volume** — Maguire, Eleanor A., David G. Gadian, Ingrid S. Johnsrude, Catriona D. Good, John Ashburner, Richard S. J. Frackowiak, and Christopher D. Frith. 'Navigation-Related Structural Change in the Hippocampi of Taxi Drivers.' *Proceedings of the National Academy of Sciences* 97, no. 8 (2000): 4398–4403. https://doi.org/10.1073/pnas.070039597. Follow-up studies confirming and extending the findings include Maguire, Woollett, and Spiers, *Hippocampus* (2006). The observation in the text that 'In 2011, a team of neuroscientists at McGill University published a study' is a simplification — the research is associated primarily with University College London (Maguire et al.) rather than McGill; McGill's Brenda Milner communicated the paper to PNAS. The 2011 GPS follow-up research is Woollett, Katherine, and Eleanor A. Maguire. 'Acquiring 'the Knowledge' of London's Layout Drives Structural Brain Changes.' *Current Biology* 21, no. 24 (2011): 2109–14.

34. **Aviation skill atrophy and manual flying requirements** — The pattern of automation-induced manual flying skill degradation is documented in multiple NTSB and EASA accident investigations. The regulatory response — mandated periodic manual flying — is reflected in FAA Advisory Circular AC 120-111 (2016), 'Upset Prevention and Recovery Training,' which addresses manual handling skill maintenance. See also: Parasuraman, R., and V. Riley. 'Humans and Automation: Use, Misuse, Disuse, Abuse.' *Human Factors* 39, no. 2 (1997): 230–53.

35. **GPS use and spatial memory degradation** — Dahmani, Louisa, and Véronique D. Bohbot. 'Habitual Use of GPS Negatively Impacts Spatial Memory During Self-Guided Navigation.' *Scientific Reports* 10 (2020): 6310. https://doi.org/10.1038/s41598-020-62877-0.

36. **Cognitive offloading** — Risko, Evan F., and Sam D. Gilbert. 'Cognitive Offloading.' *Trends in Cognitive Sciences* 20, no. 9 (2016): 676–88. The distinction between offloading that extends vs. replaces capability is the author's synthesis; the underlying concept of distributed cognition draws on Clark, Andy, and David Chalmers. 'The Extended Mind.' *Analysis* 58, no. 1 (1998): 7–19.

Chapter 8: Control, Consent, and the Illusion of Choice

37. **Ellen Langer / illusion of control** — Langer, Ellen J. 'The Illusion of Control.' *Journal of Personality and Social Psychology* 32, no. 2 (1975): 311–28. https://doi.org/10.1037/0022-3514.32.2.311. The lottery ticket experiment described in the text is Study 2 of this paper.

38. **FTC 2022 commercial surveillance report** — Federal Trade Commission. *Commercial Surveillance and Data Security: Advanced Notice of Proposed Rulemaking*, August 2022. The finding that significant user controls were non-functional or misleadingly designed is reflected throughout the commission's documentation of the proceeding.

39. **76 work days to read privacy policies** — McDonald, Aleecia M., and Lorrie Faith Cranor. 'The Cost of Reading Privacy Policies.' *I/S: A Journal of Law and Policy for the Information Society* 4, no. 3 (2008): 543–68. The paper is available at lorrie.cranor.org/pubs/readingPolicyCost-authorDraft.pdf.

40. **Nudge / libertarian paternalism** — Thaler, Richard H., and Cass R. Sunstein. *Nudge: Improving Decisions About Health, Wealth, and Happiness.* Yale University Press, 2008.

41. **Organ donation default settings** — Johnson, Eric J., and Daniel Goldstein. 'Do Defaults Save Lives?' *Science* 302, no. 5649 (2003): 1338–39. The dramatically higher donation rates in opt-out countries is well documented in this and subsequent studies.

42. **The Paradox of Choice** — Schwartz, Barry. *The Paradox of Choice: Why More Is Less.* HarperCollins, 2004.

43. **Cookie banner consent rates** — See note 27.

Chapter 9: Designing for the Human in the Loop

44. **Douglas Engelbart / Mother of All Demos** — Engelbart, Douglas C. 'Augmenting Human Intellect: A Conceptual Framework.' Stanford Research Institute, 1962. The public demonstration took place December 9, 1968, at the Fall Joint Computer Conference in San Francisco. For a full account see: Markoff, John. *What the Dormouse Said: How the Sixties Counterculture Shaped the Personal Computer Industry.* Viking, 2005.

45. **USDS passport team — 'Passport to Success'** — See note 9. Processing time reduction to one-third cited from Secretary of State Blinken's statement, September 18, 2024: 'passports on average being processed in roughly one-third the time as at the same point last summer.' 25,000 applications in first two weeks cited from Mintz.com: 'U.S. Passport Processing, Pilot Program for Online U.S. Passport Renewals,' January 9, 2023.

46. **UK passport service assessment — 'Passport to Failure'** — See note 10. The specific finding about the service manager's limited control over the backend is quoted from the published GDS assessment.

47. **Complementarity** — The complementarity principle as applied to human-AI systems draws on: Kamar, Ece. 'Directions in Hybrid Intelligence: Complementing AI Systems with Human Intelligence.' *Proceedings of IJCAI*, 2016; and Shneiderman, Ben. *Human-Centered AI* (see note 8). The specific application to UX design is the author's own synthesis.

Interlude: Design Theory 47

48. **Design Theory 47** — Personal memoir; no external citation. The Gerrit Rietveld Academie (Amsterdam) is a real institution, established 1924. The author studied design there in the 1980s.

Chapter 10: The Designer's Responsibility

49. **Joseph Rotblat** — Rotblat, Joseph, and Dario Füchs. *A World Without the Bomb: The Proceedings of the 1994 Pugwash Workshop.* World Scientific, 1995. Rotblat's departure from the Manhattan Project and subsequent peace work is documented in: Brown, Andrew. *The Neutron and the Bomb: A Biography of Sir James Chadwick.* Oxford University Press, 1997, and in Rotblat's Nobel lecture: Rotblat, Joseph. 'Remember Your Humanity.' Nobel Peace Prize lecture, December 10, 1995. Available at nobelprize.org.

50. **Artifacts have politics** — See note 20. Robert Moses's low bridges, referenced in the text, are Winner's canonical example from the same paper.

51. **UXPA Code of Professional Conduct** — User Experience Professionals Association. *UXPA Code of Professional Conduct.* Available at uxpa.org/uxpa-code-of-professional-conduct/.

52. **ACM Code of Ethics** — Association for Computing Machinery. *ACM Code of Ethics and Professional Conduct.* 2018 edition. Available at acm.org/code-of-ethics.

Further Reading

The following works are recommended for readers who want to go deeper on the book's key themes.

User Experience and Design

53. **Krug, Steve.** *Don't Make Me Think: A Common Sense Approach to Web Usability.* 3rd ed. New Riders, 2014.

54. **Cooper, Alan.** *The Inmates Are Running the Asylum: Why High-Tech Products Drive Us Crazy and How to Restore the Sanity.* Sams Publishing, 1999.

55. **Shneiderman, Ben.** *Human-Centered AI.* Oxford University Press, 2022.

56. **Norman, Don.** *The Design of Everyday Things.* Revised and expanded edition. Basic Books, 2013.

57. **Wachter-Boettcher, Sara.** *Technically Wrong: Sexist Apps, Biased Algorithms, and Other Threats of Toxic Tech.* Norton, 2017.

58. **Monteiro, Mike.** *Ruined by Design: How Designers Destroyed the World, and What We Can Do to Fix It.* Mule Books, 2019.

AI, Automation, and Society

59. **O'Neil, Cathy.** *Weapons of Math Destruction: How Big Data Increases Inequality and Threatens Democracy.* Crown, 2016.

60. **Pariser, Eli.** *The Filter Bubble: What the Internet Is Hiding from You.* Penguin Press, 2011.

61. **Noble, Safiya Umoja.** *Algorithms of Oppression: How Search Engines Reinforce Racism.* NYU Press, 2018.

62. **Zuboff, Shoshana.** *The Age of Surveillance Capitalism: The Fight for a Human Future at the New Frontier of Power.* PublicAffairs, 2019.

63. **Crawford, Kate.** *Atlas of AI: Power, Politics, and the Planetary Costs of Artificial Intelligence.* Yale University Press, 2021.

Behavioral Economics and Choice

64. **Kahneman, Daniel.** *Thinking, Fast and Slow.* Farrar, Straus and Giroux, 2011.

65. **Thaler, Richard H., and Cass R. Sunstein.** *Nudge: Improving Decisions About Health, Wealth, and Happiness.* Yale University Press, 2008.

66. **Schwartz, Barry.** *The Paradox of Choice: Why More Is Less.* HarperCollins, 2004.

Technology, Politics, and Ethics

67. **Winner, Langdon.** 'Do Artifacts Have Politics?' *Daedalus* 109, no. 1 (1980): 121–36.

68. **Benjamin, Ruha.** *Race After Technology: Abolitionist Tools for the New Jim Code.* Polity, 2019.

69. **Lanier, Jaron.** *You Are Not a Gadget: A Manifesto.* Knopf, 2010.

70. **Eubanks, Virginia.** *Automating Inequality: How High-Tech Tools Profile, Police, and Punish the Poor.* St. Martin's Press, 2018.

A Note on AI Assistance in Writing This Book

It would be inconsistent, dishonest, — and frankly ironic — to write a book about AI and user experience without disclosing how AI tools were used in its production.

Claude (Anthropic) was used as a collaborative writing partner throughout this project: for drafting, structuring, editing, and developing arguments. The ideas, professional experience, personal anecdotes, and fundamental argument of the book are the author's own.
The AI contributed to editing the prose for consistency and stopping the author going off on a rant. Boy, do I have problems sticking to the subject matter.

All factual claims were independently verified by the author. Please let me know if anything is off.

This collaboration was itself an exercise in the book's central argument. The tool was most useful when it was transparent about its limitations, easy to override, and willing to be corrected. On at least one occasion — the Sabrina Demming episode documented in Chapter 6 — it was confidently wrong in exactly the ways the book describes.

The author caught it. The seam held.

www.ingramcontent.com/pod-product-compliance
Lightning Source LLC
LaVergne TN
LVHW052251100826
845147LV00001B/19

* 9 7 8 0 9 9 1 1 5 9 2 9 1 *